ADVANCES IN CELL AGING AND GERONTOLOGY
VOLUME 16

Protein Phosphorylation in Aging and Age-Related Disease

ADVANCES IN CELL AGING AND GERONTOLOGY

VOLUME 16

Protein Phosphorylation in Aging and Age-Related Disease

Volume Editor:

Mark P. Mattson,
PhD, National Institute on Aging,
NIH
Baltimore, MD
USA

2004

ELSEVIER

Amsterdam – Boston – Heidelberg – London – New York – Oxford
Paris – San Diego – San Francisco – Singapore – Sydney – Tokyo

Elsevier
The Boulevard, Langford Lane, Kidlington, Oxford OX5 1GB, UK
Radarweg 29, PO Box 211, 1000 AE Amsterdam, The Netherlands

Library of Congress Cataloging in Publication Data
A catalog record is available from the Library of Congress.

British Library Cataloguing in Publication Data
A catalogue record is available from the British Library.

ISBN 10: 0-444-51583-6
ISBN 13: 9780444515834

Transferred to digital printing in 2007
Printed and bound by CPI Antony Rowe, Eastbourne

TABLE OF CONTENTS

PREFACE

A fundamental and evolutionarily conserved mechanism for the rapid modification of protein function is phosphorylation, a process in which a phosphate group is transferred from ATP to protein substrates on serine, threonine and tyrosine residues. In this volume of Advances in Cell Aging in Gerontology experts in the area of protein phosphorylation describe the importance of changes in protein phosphorylation in the aging of various cells, organs and organisms. In the initial chapter, Peter Atadja and Paul Kwon review information concerning how changes in protein phosphorylation may modify the process of cellular senescence with the focus on phosphorylation of transcription factors. Changes in phosphorylation of proteins linked to cell proliferation transformation and apoptosis are described. Charanjit Sandhu then provides a detailed appraisal of protein phosphorylation and the regulation of cell cycle in relation to cellular senescence in aging. A very complex set of kinases and phosphatases control the cell cycle and age-related changes in signals that lead to cellular senescence appear to prominently impact upon protein phosphorylation. In the next chapter, Bulbul Chakravarti and Deb Chakravarti reveal the importance of changes in protein phosphorylation in lymphocytes in aging of the immune system. It is well recognized that immune dysfunction occurs in aging and a better understanding of the regulation of protein phosphorylation may lead to novel ways to improve immune functions and promote successful aging. Cyclic nucleotides play prominent roles in activating kinases that modify a variety of protein substrates. Ching-Shwun Lin and Tom Lue describe the role of cyclic nucleotide signaling in the modulation of vascular smooth muscle during aging. Such alterations may play a role in major diseases such as cardiovascular disease. It is now recognized that insulin signaling plays a major role in aging in a variety of species. Eduardo Rocha and colleagues describe the phosphorylation cascades involved in insulin signaling, how they are altered in aging and how this may promote dysregulation of glucose metabolism that occurs in type 2 diabetes. Tom Foster then presents an excellent review on phosphorylation and dephosphorylation of substrates in neuronal synapses and how alterations in phosphorylation-mediated signaling may lead to dysfunction of the nervous system during aging. Finally, Lit-Fui Lau and Joel Schachter describe alterations in cytoskeletal protein phosphorylation in the nervous system and how animal models have revealed central roles for cytoskeletal alteration in age-related neurodegenerative disorders. Collectively, the chapters in this volume will provide graduate students, post-docs and senior investigators with a better understanding of the range of phosphorylation reactions that are involved in aging and age-related disease.

MARK P. MATTSON, PhD

Advances in
Cell Aging and
Gerontology

Organismal aging and phosphorylation of transcription factors

Peter W. Atadja* and Paul O. Kwon

Novartis Institute for Biomedical Research, One Health Plaza, East Hanover, NJ 07936, USA.
Tel.: +1-862-778-0435; fax: +1-973-781-7578.
E-mail address: peter.atadja@pharma.novartis.com

Contents

1. Organismal aging

Aging is the totality of degenerative changes that progressively take place in an organism beginning at birth and continuing to death. Aging is also a major phenomenon that characterizes the finite life span of living organisms. Over the past four decades there has been concerted effort, at the molecular level, to understand better, the changes that take place leading to the phenomena of aging. Observations from these studies implicate aging as a genetically programmed process where the total life span of an organism is remarkably constant and species specific. For example, the maximal life span of fruit flies is approximately 2 months, while the life span of higher-ordered animals such as mice may be up to 3 years and humans

Advances in Cell Aging and Gerontology, vol. 16, 1–14
DOI: 10.1016/S1566-3124(04)16001-X

up to 120 years (Hayflick et al., 1987). Another example supporting genetic relevance of aging is illustrated by examining twins. Interestingly, the average difference in the total longevity between fraternal twins is twice as high as that of the more genetically similar identical twins (Kalman and Jarvik, 1959). Additional genetic and molecular examples may be found in diseases such as Werners syndrome and progeria where aberrations in a single gene result in accelerated aging and reduced life span of the patients (Hayflick, 1968). Lastly, studying cultured normal cells *in vitro* and cells serially transplanted *in vivo*, we learn that the cells which were derived from living organisms have a limited proliferative capacity. This limitation in cellular proliferative ability is directly proportional to the total life span or age of the organism (Hayflick and Moorhead, 1961; Hayflick, 1965, 1984a,b, 1985a–d). These results support the study of cellular events for understanding organismal aging.

2. Cellular senescence

The phenomenon whereby nontransformed cultured cells undergo a finite number of population doublings is termed cellular senescence (Hayflick and Moorhead, 1961). Over the years, researchers have drawn parallels between *in vitro* cellular senescence and organismal aging. At the molecular level, cellular senescence is characterized by changes to the normal proliferative competency and is marked by decreased ability to synthesize DNA in response to mitogenic stimuli. These replication-deficient cells have increasingly longer cell cycle and interdivision times. The increased duration of the cell cycle in senescent cells results mainly from an extended G1 phase (Grove and Cristofalo, 1978). What remains somewhat of a mystery is that even though senescent cells have lost the critical proliferative capacity, they remain viable for months in culture without replication during which time they continue to be metabolically active. Experimentation with karyologically normal cells including fibroblasts, keratinocytes, epithelial cells, endothelial cells, smooth muscle cells, glial cells, and T-lymphocytes establish that this diverse set of cells can undergo replicative senescence demonstrating the importance and universality of cellular senescence (Rheinwald and Green, 1975; Bierman, 1978; Blomquist et al., 1980; Thornton et al., 1983; Effros and Walford, 1984). In addition to higher-ordered animal cells, senescence has also been described in *Saccharomyces cerevisiae*. Furthermore in cultured cells, the age of the cell donor appeared to have a major consequence on the proliferative capacity of those cells, and this relationship was inversely proportional. Fibroblasts derived from animals with shorter life span underwent fewer population doublings than those derived from longer-lived animal species (Smith et al., 2002). At the molecular level the activity of transcription factors such as the p53 tumor suppressor has been linked to cellular senescence. The activity of p53 was found to increase when cells undergo senescence *in vitro* (Atadja et al., 1995), whereas *in vivo*, transgenic mice expressing a stable and hyperactive form of p53 aged faster than their littermates expressing wild-type p53 (Tyner et al., 2002). Experimental observations in aging and cellular senescence provide a strong connection between these two events. Therefore, cellular senescence has been used as an *in vitro* model to study the molecular mechanisms which may be responsible for organismal aging.

3. Genetics of aging

The molecular basis of the genetic program of aging is unclear; however, two hypotheses have been postulated to explain the phenomenon of organismal and *in vitro* senescence. One proposes the existence of an intrinsic preprogrammed genetic "blueprint" that is temporally implemented over the life span of an organism to affect the changes which manifest as aging (Hayflick, 1965 1984a,b, 1985a–d). The other hypothesis explains aging as the accumulation of random errors due to physiologically damaging agents such as free radicals, reactive oxygen species (ROS), or exogenous factors that impinge on the organism from its environment. These agents directly cause cellular and tissue damage and/or cause mutation or expression of genes that in turn produce errors beyond the repair functions of the cellular repair mechanisms (Rosenberger et al., 1991). These two hypotheses are not mutually exclusive, in that, the underlying mechanisms of both involve changes in gene expression. Among the genes whose expression change when cells undergo cellular senescence are those which regulate cell growth. Generally, genes which inhibit cell growth and proliferation are upregulated whereas growth-promoting genes are expressed at lower levels in senescent cells. Since lack of growth appears to be a major component of cellular and organismal aging, dissecting the interplay between growth regulatory signals and mechanisms of gene expression during aging will reveal the critical mechanisms underlying the aging process.

4. Protein phosphorylation

Protein phosphorylation is the major mechanism by which growth signals are transduced from the extra- and intra-cellular environment of a cell to its nucleus affecting changes in gene expression. Typically, growth signals are sensed by cell surface receptor proteins with intracellular kinase domains. These kinase domains phosphorylate cytoplasmic substrates which in turn propagate additional phosphorylation events that end in the nucleus. Transcription factors are major recipients of growth regulatory phosphorylation signals and affect changes in gene expression. Phosphorylation and dephosphorylation is the most common mechanism used particularly for rapid alterations to transcription factor activity (Hunter and Karin, 1992). Modulating the phosphorylation status can alter transcription factor function by several mechanisms including controlling length of nuclear localization time, modifying chromatin structure, regulating DNA-binding activity, targeting transcriptional regulators for degradation, and modulating protein–protein interaction within the transcriptional machinery (Whitmarsh and Davis, 2000). Among the hundreds of transcription factors that are regulated through phosphorylation, several may be critical in understanding the mechanism behind the aging process. Reviewed below is the current knowledge of phosphorylation-regulated signals that alter transcription factors which may be responsible for differential expression of genes during aging. In this review, the effects of phosphorylation changes on the functions of serum response factor (SRF), p62TCF, tumor suppressors (p53 and Rb), and NF-κB are examined in greater detail.

5. The role of serum response factor in senescence

5.1. SRF function

SRF is a transcription factor that integrates cellular mitogenic signals to activate the expression of immediate early genes that mediate cell cycle progression and entry into DNA synthesis. One major characteristic of senescent cells is their inability to traverse the G1/S boundary of the cell cycle and synthesize DNA in response to mitogenic stimuli. Since molecules that regulate DNA synthesis are mostly integrated in the G1 phase of the cell cycle, studying molecular mechanisms that regulate entry into and/or progression through G1 may be most relevant for inhibiting DNA synthesis in senescent cells (Chen, 1997; Dannenberg et al., 2000; Dulic et al., 2000). As cells are stimulated to proliferate, genes belonging to the group of early response genes such as **c-fos** and **egr-1** are rapidly induced. These genes therefore function as a convergence point for many growth signal cascades and are transcription factors themselves thought to induce the expression of "delayed genes" that play a critical role in later phases of the cell cycle enabling DNA synthesis. Attenuated levels of **c-fos** and **egr-1** found in senescent cells suggest that repression of these proteins may be involved in maintaining the growth-arrested state of cellular senescence (Seshadri and Campisi, 1990; Riabowol et al., 1992; Meyyappan et al., 1999). Both **c-fos** and **egr-1** promoters have *cis* elements that appear to be important for transcriptional activation. At least one important regulatory element is the serum response element (SRE), the specific target of SRF (Treisman, 1986; Fisch et al., 1987; Greenberg et al., 1987; Gilman, 1988).

5.2. Immediate early genes (c-fos and egr-1)

The mechanism behind the growth-arrested state in cellular senescence might be due to differential regulation of genes such as **c-fos** and **egr-1**, which are growth factor-stimulated immediate early genes that are also believed to link mitogenic signals to cell cycle progression. **c-fos** is the cellular homolog of the transforming oncogene from the FBJ murine osteosarcoma virus and its protein product is thought to play a critical role in G1 progression. Decreased expression of the **c-fos** gene observed in senescent cells may explain their inability to synthesize DNA and consequently fail to proceed normally through cell cycle. Studies in which **c-fos** expression was inhibited by microinjection of anti-Fos antibodies or by using antisense constructs showed blocked DNA synthesis in proliferation competent fibroblasts (Holt et al., 1986; Nishikura and Murray, 1987; Riabowol et al., 1988; Kovary and Bravo, 1991). More interestingly, a limited renewal of DNA synthesis occurred in senescent WI38 cells in response to overexpressed Fos (Phillips et al., 1992).

In addition, **egr-1** is a zinc finger transcription factor discovered while searching for genes essential for growth, proliferation, and differentiation. Serum stimulation of cells results in approximately 20-fold induction of **egr-1** mRNA that rapidly decays 2–3 h thereafter (Sukhatme et al., 1988; Sukhatme, 1990). Meyyappan et al. examined the expression and activity of **egr-1** in senescent fibroblasts and found several-fold lower expression levels in late passage cells compared to early passage cells. Initial transcription factor binding analysis of the **c-fos** and **egr-1** promoters revealed SRF as an important regulatory element. A dramatic decrease in the region of the SRE was detected on the **c-fos** promoter of senescent cells, whereas a marked decrease of SRF binding to the SRE on the **egr-1** promoter was observed in nuclear extracts prepared from senescent fibroblasts (Rittling et al., 1986; Chang and Chen, 1988; Seshadri and Campisi, 1990; Riabowol et al., 1992; Atadja et al., 1994; Atadja, Ph.D. dissertation, 1995; Goyns et al., 1998; Meyyappan et al., 1999). These results confirm abnormal functioning of SRF in senescent cells and suggest a common mechanism for the down-regulation of mitogen-responsive genes during cellular senescence.

5.3. SRF in the ternary complex

Since the above evidence implicates SRF as an important element in progression of senescence, there has been much effort to elucidate mechanisms that change SRF activity. Much of the focus has been to further characterize changes to the SRE–protein complex in senescent cells utilizing the **c-fos** promoter as a model. SRF is a 67-kDa phosphoprotein which binds to the SRE as a homodimer and integrates cellular mitogenic signals to activate expression of immediate early genes. The 20 bp sequence of dyad symmetry that the SRF binds is located 300 bp upstream of the **c-fos** transcription initiation site and is both necessary and sufficient to confer the normal transient kinetics of transcription in response to mitogens by the intact **c-fos** promoter (Greenberg and Ziff, 1984; Rivera and Greenberg, 1990). Purified SRF was found to be phosphorylated mainly in the amino-terminal part of the protein by several kinases, including Casein kinase II, DNA-dependent protein kinase, p90rsk, and MAP kinases (Manak and Prywes, 1991; Misra et al., 1991; Janknecht et al., 1992; Marais et al., 1992, 1993; Heidenreich et al., 1999). A number of studies were conducted to assess the effect of phosphorylation on SRF activity. Misra et al. showed that SRF is transiently synthesized following serum stimulation and rapidly transported into the nucleus where it is increasingly modified by phosphorylation (Manak and Prywes, 1991; Misra et al., 1991). Other studies indicate that phosphorylation of SRF by Casein kinase II does not affect its steady-state binding of the SRE per se, but it rather increases the on/off rates of association with the **c-fos** promoter (Janknecht et al., 1992; Marais et al., 1992). SRF activity is further controlled by its association with p62TCF group of factors and this complex is termed the ternary complex. p62TCF (which is encoded by Elk-1, Sap1, and Sap2) contains a transcriptional activation domain that is activated by MAP kinases (Gille et al., 1992; Whitmarsh et al., 1995; Wang and Prywes, 2000). Phosphorylation of p62TCF has been associated with an increase in ternary complex formation upon growth factor stimulation or exposure of cells to stress (Malik et al., 1991; Gille et al., 1992, 1995; Thompson et al., 1994). Taken together, transcription factor phosphorylation may be a major mechanism by which the expression of protooncogene **c-fos** is activated in response to growth factors.

Experimental techniques, such as electrophoretic mobility assays, identified four specific ternary complexes formed on the wild-type SRE in proliferation-competent fibroblasts. However, during late passages of fibroblast cells three forms of the ternary complex were undetectable. Antibody supershifting experiments revealed that the missing complexes contained SRF and resulted from decreased formation of the ternary complex, not from reduced protein levels (Atadja et al., 1994). Further investigation of affinity-purified SRF showed two forms with slightly different mobility characteristics, suggesting differential posttranslational modification. To identify whether the different migration pattern was due to differing phosphorylation status of SRF, early and late passage cells were labeled with 32P orthophosphate, and SRF was immunoprecipated from them. Intriguingly, more 32P-labeled SRF was immunoprecipitated from late passage cells (Atadja et al., 1994). Moreover, considering that the phosphorylation status of transcription factors affect DNA-binding, mobility shift assays were performed with nuclear extracts from early and late passage cells in the presence or absence of sodium fluoride, a phosphatase inhibitor. Results from Atadja et al. (1994) showed that nuclear extracts from senescent cells in the absence of phosphatase inhibition bound the SRE in equivalent amounts to extracts from early passage cells. This result indicates that the presence of protein phosphatases in senescent nuclear extracts, when not inhibited, would reduce the hyperphosphorylation state of SRF to enable efficient binding to the SRE. Thus, SRF hyperphosphorylation is likely the cause of its inability to bind the **c-fos** promoter in senescent fibroblasts, leading to decreased **c-fos** expression and presumably inhibiting senescent cells from synthesizing DNA in response to growth factor stimulation.

6. P62TCF in senescent cells

The components of p62TCF which forms the ternary complex with SRF on the **c-fos** SRE is also regulated through phosphorylation by extracellular-regulated kinase (ERK) family. The diminished nucleoprotein complex formation on the SRE in senescent cells caused Tresini et al. (2001) to study the expression and activity of p62TCF proteins in late passage cells. Although p62TCF does not bind to the c-fos SRE in the absence of SRF, the p62TCF proteins were found to bind Ets elements in the promoter of the *Drosophila* E74 gene with high affinity in an autonomous manner (Shore and Sharrocks, 1995). In addition, Tresini et al. discovered that under senescent conditions p62TCF proteins bind Ets elements at a reduced level as compared to early passage cells. However, the level of Ets elements was similar in "young" and "old" cells. Further analysis of the mitogen-induced phosphorylation of p62TCF proteins showed an abundance of serine phosphorylated Elk-1, a TCF-like protein, in response to serum stimulation in early passage cells but this effect was greatly diminished in senescent cells. Decreased phosphorylation of Elk-1 correlated with increased accumulation of inactive components of the MAP kinase pathway such as unphosphorylated ERK and Mek. These findings further implicate altered phosphorylation of transcription factors in the decreased **c-fos** expression seen during cellular aging.

7. Tumor suppressors (p53 and Rb) in cellular senescence

7.1. Regulation of p53

Evidence is accumulating that replicative senescence might be a mechanism for suppressing tumorigenesis in aging organisms (Atadja et al., 1995; Krtolica et al., 2001). As such, the activities of tumor-suppressor proteins in cellular aging have been studied (Atadja et al., 1995; Schmitt et al., 2002). Two transcription factor tumor-suppressor proteins that are intricately involved in regulating progression through the G1/S phase of the cell cycle are p53 and Rb. The p53 tumor-suppressor protein functions as a negative growth regulator by binding in a sequence-specific manner and modulating the expression of other growth regulatory genes such as p21, mdm2, and GADD45 (Kastan et al., 1992; Momand et al., 1992; el-Deiry et al., 1993). The p53-activated gene p21 encodes a protein that inhibits the activity of G1 cyclin-dependent kinases that are required for progression through the G1/S checkpoint. The p21 gene was originally identified as a gene whose expression was dramatically increased in senescent human diploid fibroblasts and whose product inhibited DNA synthesis and cell growth when overexpressed in normal early passage cells (Noda et al., 2001). Since expression of the p53-regulated p21 gene is increased in senescent cells, studies were conducted to examine activity of the p53 transcription factor during cellular aging (Atadja et al., 1995). Both DNA-binding and transcriptional activities of p53 were enhanced as cells aged *in vitro*; however, the amount of p53 protein in early and late passage cells remained constant (Atadja et al., 1995). Other studies have also shown increased p53 activity as cells undergo cellular senescence (Bond et al., 1996; Vaziri et al., 1997). More recently, heterozygotic mice carrying a p53 mutation that enhanced the stability of the wild-type protein showed shorter life span and faster aging (Cao et al., 2003). These mice also developed fewer tumors, linking organismal aging to tumor suppression. A role for p53 phosphorylation in oncogenic ras-induced cellular senescence has been described (Bischoff et al., 2002). Ectopic expression of oncogenic ras induces the expression and accumulation of the leukemic protein PML (Pearson et al., 2000), leading to a premature senescence in human diploid fibroblasts. PML-induced senescence involves stabilization and activation of p53 through phosphorylation at Ser46 and acetylation at Lys382, and this occurs independent of telomerase. Thus, a role for p53 phosphorylation is suggested in at least one model of cellular senescence.

7.2. The role of Rb

The retinoblastoma protein (Rb) is a well-characterized protein with known transcriptional repression activity that is differentially phosphorylated during the cell cycle (Hatakeyama and Weinberg, 1995; Chau and Wang, 2003). In replication-competent cells, the Rb protein is hyperphosphorylated in response to mitogen stimulation just prior to the G1/S boundary of the cell cycle. The repressive properties are lost allowing induction of growth-promoting genes and progression into S-phase. On the other hand, Rb is relatively hypophosphorylated during quiescence. In this form it functions as a transcriptional repressor of growth-promoting genes, such as E2F, and is thought to play a role in maintaining the quiescent state. Interestingly, a number of studies have shown that Rb fails to become phosphorylated in response to serum stimulation in senescent fibroblasts (Stein et al., 1991; Barret, 1991; Riabowol et al., 1992). This accumulation of unphosphorylated Rb protein, a state similar to quiescence, may be responsible for inhibiting cellular proliferation.

Further evidence suggesting a role for Rb in senescence comes from the fact that Rb interacts with specific domains of certain viral oncoproteins including the SV40 T-antigen, the adenovirus E1A protein, and the E7 protein of the human papilloma virus. Each of these viral proteins are capable of inducing DNA synthesis in senescent cells and in some cases conferring an immortal phenotype (Buchkovich et al., 1989; DeCaprio et al., 1989). SV40 T-antigen binds to the unphosphorylated (active) form of Rb and blocks its function as a negative regulator of transcription. In a study to determine whether Rb plays a role in growth arrest associated with cellular senescence, Campisi (1992) microinjected SV40 T-antigen into senescent human fibroblasts and found the resumption of a limited DNA synthesis. In contrast, microinjection of a mutant form of SV40 T-antigen that is not able to bind Rb did not induce DNA synthesis in senescent cells. These results suggest that overcoming Rb-regulated mechanisms might be one way by which the SV40 T-antigen induces DNA synthesis in senescent cells. Hara et al. also addressed the question by introducing Rb antisense oligonucleotides into senescent fibroblasts and observed an extension of the normal life span by approximately 10 population doublings (Hara et al., 1991). The aforementioned observations suggest an important role of the Rb tumor suppressor as a differentially phosphorylated transcriptional suppressor in cellular senescence.

7.3. Cross-talk between the p53 and Rb pathway

A closer examination of both p53 and Rb pathways may explain the role of p21 in cell cycle regulation as well as senescence. Rb is phosphorylated by cyclin-dependent kinases (cdk) in the G1 phase of the cell cycle. The expression of the p53 inducible protein p21, an inhibitor of cyclin–cdk activity, is upregulated in senescent cells. Thus, p21-mediated inhibition of cyclin–cdk activity in senescent cells is a possible reason for decreased Rb phosphorylation. In this function, p21 brings together two major tumor-suppression pathways (Rb and p53) to block DNA synthesis and proliferation in senescent cells.

8. NF-κB activation in aging

There is growing evidence that most age-related changes are causally linked to deleterious oxygen species (Yu, 1994; Martin et al., 1996; Levine, 1998; Fukagawa, 1999) and that the deleterious processes associated with age might evolve through oxidatively stressed proinflammatory processes (Chung et al., 2000). Calorific restriction, the only established anti-aging model, has been shown to prolong organismal life span and functions through suppression of age-related oxidative stress (Sohal and Weindruch, 1996; Yu, 1996; Yu and Chung, 2001) and/or by increasing cellular stress resistance (Mattson et al., 2003). A recent report linked activation of NF-κB subsequent to increased degradation of IκBα during age-related oxidative and inflammatory responses (Kim et al., 2000). NF-κB, a transcription factor that is sensitive to oxidative stress, produces inflammatory responses. NF-κB interacts with inhibitor IκB proteins resulting in cytoplasmic sequestration leading to attenuated DNA-binding and transcriptional activation. Phosphorylation of IκB proteins results in their degradation allowing translocation of NF-κB into the nucleus which leads to transcription of inflammatory genes. Kim et al. (2000) have reported the correlation of NF-κB activation and age-related oxidative stress. This process appears to result from increased activity of IκB kinase (IKK), causing rapid ubiquitination and degradation of IκBα and IκBβ (Kim et al., 2000, 2002). Interestingly, Kim et al. showed that calorific restriction inhibits IKK activation, downregulating NF-κB activity as evidenced by increased level of the NF-κB–IκB complex in the cytoplasm (Kim et al., 2002). These findings implicate NF-κB signaling pathways and the significance of phosphorylation signals as regulatory mechanisms of aging.

9. Conclusions

Aging is a universal phenomena for all living organisms and helps define our individual understanding of mortality and finality. Philosophically and biologically, it has fascinated us as we come to terms with prospects of growing old. In the recent few decades significant scientific advances have been made to understand this process, yet much more research is needed to clearly comprehend this intricate and complex process (Fig. 1). Experimental data suggest that both genetic and epigenetic factors contribute to organismal aging. Moreover, there is strong evidence linking molecular mechanisms in cellular senescence with aging, and deciphering some of these mechanisms revealed differential gene expression levels between "old" and "young" cells. This variation in gene expression was found to result from altered transcription factor functions arising from differential phosphorylation states. In this review several important transcription factors thought to regulate cell cycle control and mediate responses to internal and external stimuli are discussed. These factors include SRF, p62TCF, tumor suppressors, and NF-κB. Considering that these transcription factors integrate and effect different signaling pathways, the consequence of aberrant phosphorylation states leading to

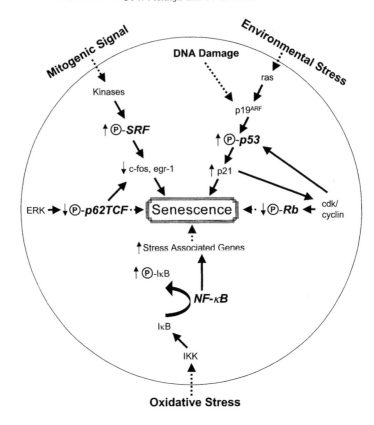

Fig. 1.

misregulation and/or disruption of the "normal" signaling cascade is significant.
Therefore, unveiling the cellular activities of these factors is critical in understanding
the aging process and other diseases, where stringent transcriptional regulation needs
to be maintained.

References

Atadja, P.W., 1995 Aug. Molecular Analyses of Cellular Replicative Senescence. The University of
    Calgary.
Atadja, P.W., Stringer, K.F., Riabowol, K.T., 1994 Jul. Loss of serum response element-binding activity
    and hyperphosphorylation of serum response factor during cellular aging. Mol. Cell. Biol. 14(7),
    4991–4999.
Atadja, P., Wong, H., Garkavtsev, I., Veillette, C., Riabowol, K., 1995 Aug 29. Increased activity of p53 in
    senescing fibroblasts. Proc. Natl. Acad. Sci. USA 92(18), 8348–8352.
Barret, J.C., 1991. Age-related changes in the cell kinetics of rat foot epidermis. Cell Prolif. 24, 239–240.
Bierman, E.L., 1978 Nov. The effect of donor age on the in vitro life span of cultured human arterial
    smooth-muscle cells. In Vitro 14(11), 951–955.
Bischoff, F.Z., Heard, M., Simpson, J.L., 2002 May–Jun. Somatic DNA alterations in endometriosis: high
    frequency of chromosome 17 and p53 loss in late-stage endometriosis. J. Reprod. Immunol. 55(1–2),
    49–64.

Blomquist, E., Westermark, B., Ponten, J., 1980 Feb. Ageing of human glial cells in culture: increase in the fraction of non-dividers as demonstrated by a minicloning technique. Mech. Ageing Dev. 12(2), 173–182.

Bond, J., Haughton, M., Blaydes, J., Gire, V., Wynford-Thomas, D., Wyllie, F., 1996 Nov 21. Evidence that transcriptional activation by p53 plays a direct role in the induction of cellular senescence. Oncogene 13(10), 2097–2104.

Buchkovich, K., Duffy, L.A., Harlow, E., 1989 Sep 22. The retinoblastoma protein is phosphorylated during specific phases of the cell cycle. Cell 58(6), 1097–1105.

Campisi, J., 1992 Nov 21. Gene expression in quiescent and senescent fibroblasts. Ann. N.Y. Acad. Sci. 663, 195–201.

Cao, L., Li, W., Kim, S., Brodie, S.G., Deng, C.X., 2003 Jan 15. Senescence, aging, and malignant transformation mediated by p53 in mice lacking the Brca1 full-length isoform. Genes Dev. 17(2), 201–213.

Chang, Z.F., Chen, K.Y., 1988 Aug 15. Regulation of ornithine decarboxylase and other cell cycle-dependent genes during senescence of IMR-90 human diploid fibroblasts. J. Biol. Chem. 263(23), 11431–11435.

Chau, B.N., Wang, J.Y., 2003 Feb. Coordinated regulation of life and death by RB. Nat. Rev. Cancer 3(2), 130–138.

Chen, K.Y., 1997 Sep 1. Transcription factors and the down-regulation of G1/S boundary genes in human diploid fibroblasts during senescence. Front. Biosci. 2, d417–d426.

Chung, Y.H., Shin, C., Kim, M.J., Lee, B., Park, K.H., Cha, C.I., 2000 Dec 1. Immunocytochemical study on the distribution of p53 in the hippocampus and cerebellum of the aged rat. Brain Res. 885(1), 137–141.

Dannenberg, J.H., van Rossum, A., Schuijff, L., te Riele, H., 2000 Dec 1. Ablation of the retinoblastoma gene family deregulates G(1) control causing immortalization and increased cell turnover under growth-restricting conditions. Genes Dev. 14(23), 3051–3064.

DeCaprio, J.A., Ludlow, J.W., Lynch, D., Furukawa, Y., Griffin, J., Piwnica-Worms, H., Huang, C.M., Livingston, D.M., 1989 Sep 22. The product of the retinoblastoma susceptibility gene has properties of a cell cycle regulatory element. Cell 58(6), 1085–1095.

Dulic, V., Beney, G.E., Frebourg, G., Drullinger, L.F., Stein, G.H., 2000 Sep. Uncoupling between phenotypic senescence and cell cycle arrest in aging p21-deficient fibroblasts. Mol. Cell. Biol. 20(18), 6741–6754.

Effros, R.B., Walford, R.L., 1984 Oct 15. The effect of age on the antigen-presenting mechanism in limiting dilution precursor cell frequency analysis. Cell Immunol. 88(2), 531–539.

el-Deiry, W.S., Tokino, T., Velculescu, V.E., Levy, D.B., Parsons, R., Trent, J.M., Lin, D., Mercer, W.E., Kinzler, K.W., Vogelstein, B., 1993 Nov 19. WAF1, a potential mediator of p53 tumor suppression. Cell 75(4), 817–825.

Fisch, T.M., Prywes, R., Roeder, R.G., 1987 Oct. c-fos sequence necessary for basal expression and induction by epidermal growth factor, 12-O-tetradecanoyl phorbol-13-acetate and the calcium ionophore. Mol. Cell. Biol. 7(10), 3490–3502.

Fukagawa, N.K., 1999 Dec. Aging: is oxidative stress a marker or is it causal? Proc. Soc. Exp. Biol. Med. 222(3), 293–298.

Gille, H., Sharrocks, A.D., Shaw, P.E., 1992 Jul 30. Phosphorylation of transcription factor p62TCF by MAP kinase stimulates ternary complex formation at c-fos promoter. Nature 358(6385), 414–417.

Gilman, M.Z., 1988 Apr. The c-fos serum response element responds to protein kinase C-dependent and -independent signals but not to cyclic AMP. Genes Dev. 2(4), 394–402.

Goyns, M.H., Charlton, M.A., Dunford, J.E., Lavery, W.L., Merry, B.J., Salehi, M., Simoes, D.C., 1998 Mar 16. Differential display analysis of gene expression indicates that age-related changes are restricted to a small cohort of genes. Mech. Ageing Dev. 101(1–2), 73–90.

Greenberg, M.E., Ziff, E.B., 1984 Oct 4–10. Stimulation of 3T3 cells induces transcription of the c-fos proto-oncogene. Nature 311(5985), 433–438.

Greenberg, M.E., Siegfried, Z., Ziff, E.B., 1987 Mar. Mutation of the c-fos gene dyad symmetry element inhibits serum inducibility of transcription in vivo and the nuclear regulatory factor binding in vitro. Mol. Cell. Biol. 7(3), 1217–1225.

Grove, G.L., Cristofalo, V.J., 1978 Mar. Transition probability model and aging human diploid cell cultures: a reply to Smith (1977). Cell Biol. Int. Rep. 2(2), 185–188.

Hara, E., Tsurui, H., Shinozaki, A., Nakada, S., Oda, K., 1991 Aug 30. Cooperative effect of antisense-Rb and antisense-p53 oligomers on the extension of life span in human diploid fibroblasts, TIG-1. Biochem. Biophys. Res. Commun. 179(1), 528–534.

Hatakeyama, M., Weinberg, R.A., 1995. The role of RB in cell cycle control. Prog. Cell Cycle Res. 1, 9–19.

Hayflick, L., 1965 Jun. Tissue cultures and mycoplasmas. Tex. Rep. Biol. Med. 23(Suppl. 1), 285.

Hayflick, L., 1968 Mar. Human cells and aging. Sci. Am. 218(3), 32–37.

Hayflick, L., 1984a Dec. Intracellular determinants of cell aging. Mech. Ageing Dev. 28(2–3), 177–185.

Hayflick, L., 1984b Mar. When does aging begin? Res. Aging 6(1), 99–103.

Hayflick, L., 1985a Feb. The cell biology of aging. Clin. Geriatr. Med. 1(1), 15–27.

Hayflick, L., 1985b. Future directions in aging research. Basic Life Sci. 35, 447–460.

Hayflick, L., 1985c. The aging process: current theories. Drug Nutr. Interact. 4(1–2), 13–33.

Hayflick, L., 1985d. Theories of biological aging. Exp. Gerontol. 20(3–4), 145–159.

Hayflick, L., Moorhead, P.S., 1961. The serial civilization of human diploid cell strains. Exp. Cell Res. 25, 585–621.

Hayflick, L., Plotkin, S., Stevenson, R.E., 1987. History of the acceptance of human diploid cell strains as substrates for human virus vaccine manufacture. Dev. Biol. Stand. 68, 9–17.

Heidenreich, O., Neininger, A., Schratt, G., Zinck, R., Cahill, M.A., Engel, K., Kotlyarov, A., Kraft, R., Kostka, S., Gaestel, M., Nordheim, A., 1999 May 14. MAPKAP kinase 2 phosphorylates serum response factor in vitro and in vivo. J. Biol. Chem. 274(20), 14434–14443.

Holt, J.T., Gopal, T.V., Moulton, A.D., Nienhuis, A.W., 1986 Jul. Inducible production of c-fos antisense RNA inhibits 3T3 cell proliferation. Proc. Natl. Acad. Sci. USA 83(13), 4794–4798.

Hunter, T., Karin, M., 1992 Aug 7. The regulation of transcription by phosphorylation. Cell 70(3), 375–387.

Janknecht, R., Hipskind, R.A., Houthaeve, T., Nordheim, A., Stunnenberg, H.G., 1992 Mar. Identification of multiple SRF N-terminal phosphorylation sites affecting DNA binding properties. EMBO J. 11(3), 1045–1054.

Kalman, E.J., Jarvik, L.F., 1959. Individual differences in constitution and genetic background. Handbook of Aging and the Individual 30–64.

Kastan, M.B., Zhan, Q., el-Deiry, W.S., Carrier, F., Jacks, T., Walsh, W.V., Plunkett, B.S., Vogelstein, B., Fornace, A.J., Jr., 1992 Nov 13. A mammalian cell cycle checkpoint pathway utilizing p53 and GADD45 is defective in ataxia-telangiectasia. Cell 71(4), 587–597.

Kim, H.J., Kim, K.W., Yu, B.P., Chung, H.Y., 2000 Mar 1. The effect of age on cyclooxygenase-2 gene expression: NF-kappaB activation and IkappaBalpha degradation. Free Radic. Biol. Med. 28(5), 683–692.

Kim, H.J., Yu, B.P., Chung, H.Y., 2002 May 15. Molecular exploration of age-related NF-kappaB/IKK downregulation by calorie restriction in rat kidney. Free Radic. Biol. Med. 32(10), 991–1005.

Kovary, K., Bravo, R., 1991 Sep. The jun and fos protein families are both required for cell cycle progression in fibroblasts. Mol. Cell. Biol. 11(9), 4466–4472.

Krtolica, A., Parrinello, S., Lockett, S., Desprez, P.Y., Campisi, J., 2001 Oct 9. Senescent fibroblasts promote epithelial cell growth and tumorigenesis: a link between cancer and aging. Proc. Natl. Acad. Sci. USA 98(21), 12072–12077.

Levine, R.L., 1998 Apr. Oxidative stress and aging. Aging 10(2), 151.

Malik, R.K., Roe, M.W., Blackshear, P.J., 1991 May. Epidermal growth factor and other mitogens induce binding of a protein complex to the c-fos serum response element in human astrocytoma and other cells. J. Biol. Chem. 5, 266(13), 8576–8582.

Manak, J.R., Prywes, R., 1991 Jul. Mutation of serum response factor phosphorylation sites and the mechanism by which its DNA-binding activity is increased by casein kinase II. Mol. Cell. Biol. 11(7), 3652–3659.

Marais, R.M., Hsuan, J.J., McGuigan, C., Wynne, J., Treisman, R., 1992 Jan. Casein kinase II phosphorylation increases the rate of serum response factor-binding site exchange. EMBO J. 11(1), 97–105.

Marais, R., Wynne, J., Treisman, R., 1993 Apr 23. The SRF accessory protein Elk-1 contains a growth factor-regulated transcriptional activation domain. Cell 73(2), 381–393.

Martin, G.M., Austad, S.N., Johnson, T.E., 1996 May. Genetic analysis of ageing: role of oxidative damage and environmental stresses. Nat. Genet. 13(1), 25–34.

Mattson, M.P., Duan, W., Guo, Z., 2003. Meal size and frequency affect neuronal plasticity and vulnerability to disease: cellular and molecular mechanisms. J. Neurochem. 84, 417–431.

Meyyappan, M., Wheaton, K., Riabowol, K.T., 1999 Apr. Decreased expression and activity of the immediate-early growth response (Egr-1) gene product during cellular senescence. J. Cell Physiol. 179(1), 29–39.

Misra, R.P., Rivera, V.M., Wang, J.M., Fan, P.D., Greenberg, M.E., 1991 Sep. The serum response factor is extensively modified by phosphorylation following its synthesis in serum-stimulated fibroblasts. Mol. Cell. Biol. 11(9), 4545–4554.

Momand, J., Zambetti, G.P., Olson, D.C., George, D., Levine, A.J., 1992 Jun 26. The mdm-2 oncogene product forms a complex with the p53 protein and inhibits p53-mediated transactivation. Cell 69(7), 1237–1245.

Nishikura, K., Murray, J.M., 1987 Feb. Antisense RNA of proto-oncogene c-fos blocks renewed growth of quiescent 3T3 cells. Mol. Cell. Biol. 7(2), 639–649.

Noda, H., Maehara, Y., Irie, K., Kakeji, Y., Yonemura, T., Sugimachi, K., 2001 Oct 1. Growth pattern and expressions of cell cycle regulator proteins p53 and p21WAF1/CIP1 in early gastric carcinoma. Cancer 92(7), 1828–1835.

Pearson, M., Carbone, R., Sebastiani, C., Cioce, M., Fagioli, M., Saito, S., Higashimoto, Y., Appella, E., Minucci, S., Pandolfi, P.P., Pelicci, P.G., 2000 Jul 13. PML regulates p53 acetylation and premature senescence induced by oncogenic Ras. Nature 406(6792), 207–210.

Phillips, P.D., Pignolo, R.J., Nishikura, K., Cristofalo, V.J., 1992 Apr. Renewed DNA synthesis in senescent WI-38 cells by expression of an inducible chimeric c-fos construct. J. Cell Physiol. 151(1), 206–212.

Rheinwald, J.G., Green, H., 1975 Nov. Serial cultivation of strains of human epidermal keratinocytes: the formation of keratinizing colonies from single cells. Cell 6(3), 331–343.

Riabowol, K.T., Vosatka, R.J., Ziff, E.B., Lamb, N.J., Feramisco, J.R., 1988 Apr. Microinjection of fcs-specific antibodies blocks DNA synthesis in fibroblast cells. Mol. Cell. Biol. 8(4), 1670–1676.

Riabowol, K., Schiff, J., Gilman, M.Z., 1992 Jan 1. Transcription factor AP-1 activity is required for initiation of DNA synthesis and is lost during cellular aging. Proc. Natl. Acad. Sci. USA 89(1), 157–161.

Rittling, S.R., Brooks, K.M., Cristofalo, V.J., Baserga, R., 1986 May. Expression of cell cycle-dependent genes in young and senescent WI-38 fibroblasts. Proc. Natl. Acad. Sci. USA 83(10), 3316–3320.

Rivera, V.M., Greenberg, M.E., 1990 Sep. Growth factor-induced gene expression: the ups and downs of c-fos regulation. New Biol. 2(9), 751–758.

Rosenberger, R.F., Gounaris, E., Kolettas, E., 1991 Feb 7. Mechanisms responsible for the limited lifespan and immortal phenotypes in cultured mammalian cells. J. Theor. Biol. 148(3), 383–392.

Schmitt, C.A., Fridman, J.S., Yang, M., Lee, S., Baranov, E., Hoffman, R.M., Lowe, S.W., 2002 May 3. A senescence program controlled by p53 and p16INK4a contributes to the outcome of cancer therapy. Cell 109(3), 335–346.

Seshadri, T., Campisi, J., 1990 Jan 12. Repression of c-fos transcription and an altered genetic program in senescent human fibroblasts. Science 247(4939), 205–209.

Shore, P., Sharrocks, A.D., 1995 Nov 25. The ETS-domain transcription factors Elk-1 and SAP-1 exhibit differential DNA binding specificities. Nucleic Acids Res. 23(22), 4698–4706.

Smith, J.R., Venable, S., Roberts, T.W., Metter, E.J., Monticone, R., Schneider, E.L., 2002 Jun. Relationship between in vivo age and in vitro aging: assessment of 669 cell cultures derived from members of the Baltimore Longitudinal Study of Aging. J. Gerontol. A. Biol. Sci. Med. Sci. 57(6), B239–B246.

Sohal, R.S., Weindruch, R., 1996 Jul 5. Oxidative stress, caloric restriction, and aging. Science 273(5271), 59–63.

Stein, G.H., Drullinger, L.F., Robetorye, R.S., Pereira-Smith, O.M., Smith, J.R., 1991 Dec 15. Senescent cells fail to express cdc2, cycA, and cycB in response to mitogen stimulation. Proc. Natl. Acad. Sci. USA 88(24), 11012–11016.

Sukhatme, V.P., 1990 Dec. Early transcriptional events in cell growth: the Egr family. J. Am. Soc. Nephrol. 1(6), 859–866.

Sukhatme, V.P., Cao, X.M., Chang, L.C., Tsai-Morris, C.H., Stamenkovich, D., Ferreira, P.C., Cohen, D.R., Edwards, S.A., Shows, T.B., Curran, T., et al., 1988 Apr 8. A zinc finger-encoding gene coregulated with c-fos during growth and differentiation, and after cellular depolarization. Cell 53(1), 37–43.

Thompson, M.J., Roe, M.W., Malik, R.K., Blackshear, P.J., 1994 Aug 19. Insulin and other growth factors induce binding of the ternary complex and a novel protein complex to the c-fos serum response element. J. Biol. Chem. 269(33), 21127–21135.

Thornton, S.C., Mueller, S.N., Levine, E.M., 1983 Nov 11. Human endothelial cells: use of heparin in cloning and long-term serial cultivation. Science 222(4624), 623–625.

Treisman, R., 1986 Aug 15. Identification of a protein-binding site that mediates transcriptional response of the c-fos gene to serum factors. Cell 46(4), 567–574.

Tresini, M., Lorenzini, A., Frisoni, L., Allen, R.G., Cristofalo, V.J., 2001 Oct 1. Lack of Elk-1 phosphorylation and dysregulation of the extracellular regulated kinase signaling pathway in senescent human fibroblast. Exp. Cell Res. 269(2), 287–300.

Tyner, S.D., Venkatachalam, S., Choi, J., Jones, S., Ghebranious, N., Igelmann, H., Lu, X., Soron, G., Cooper, B., Brayton, C., Hee Park, S., Thompson, T., Karsenty, G., Bradley, A., Donehower, L.A., 2002 Jan 3. p53 mutant mice that display early ageing-associated phenotypes. Nature 415(6867), 45–53.

Vaziri, H., West, M.D., Allsopp, R.C., Davison, T.S., Wu, Y.S., Arrowsmith, C.H., Poirier, G.G., Benchimol, S., 1997 Oct 1. ATM-dependent telomere loss in aging human diploid fibroblasts and DNA damage lead to the post-translational activation of p53 protein involving poly(ADP-ribose) polymerase. EMBO J. 16(19), 6018–6033.

Wang, Y., Prywes, R., 2000 Mar 9. Activation of the c-fos enhancer by the erk MAP kinase pathway through two sequence elements: the c-fos AP-1 and p62TCF sites. Oncogene 19(11), 1379–1385.

Whitmarsh, A.J., Davis, R.J., 2000 Aug. Regulation of transcription factor function by phosphorylation. Cell. Mol. Life Sci. 57(8–9), 1172–1183.

Whitmarsh, A.J., Shore, P., Sharrocks, A.D., Davis, R.J., 1995 Jul 21. Integration of MAP kinase signal transduction pathways at the serum response element. Science 269(5222), 403–407.

Yu, B.P., 1994 Jan. Cellular defenses against damage from reactive oxygen species. Physiol. Rev. 74(1), 139–162.

Yu, B.P., 1996. Aging and oxidative stress: modulation by dietary restriction. Free Radic. Biol. Med. 21(5), 651–668.

Yu, B.P., Chung, H.Y., 2001 Apr. Stress resistance by caloric restriction for longevity. Ann. N.Y. Acad. Sci. 928, 39–47.

Advances in
Cell Aging and
Gerontology

Phosphorylation of cell cycle proteins at senescence

Charanjit Sandhu

*Program in Proteomics and Bioinformatics, Banting and Best Department of Medical Research,
University of Toronto, Toronto, ON, Canada.*
*Correspondence address: CH Best Institute, University of Toronto, 112 College Street, Rm 402,
Toronto, ON, Canada M5G 1L6.*
E-mail address: charanjit.sandhu@utoronto.ca

Contents

1. Overview
2. Cell cycle overview
3. pRB- and E2F-mediated transcriptional control
4. Cyclins: A requirement for cdk activation
5. Cdk inhibitors regulate cdk activity
6. INK family of inhibitors
7. Cdk phosphorylation
8. Cellular senescence
9. Cell cycle, senescence, and protein phosphorylation
10. Summary and future directions

1. Overview

In the past decade, significant advances have been made in our understanding of the mechanisms that regulate cellular growth and govern cellular senescence. This includes a detailed understanding of the mitogens that stimulate growth, the receptors to which they bind, the proteins that mediate these signals into the cell, and finally, the proteins that effect these changes: the cell cycle regulators. Cell cycle arrest at senescence is widely believed to be triggered by a critical shortening of telomere length (Mathon, 2001). Protein effectors of cell cycle progression are tightly regulated by phosphorylation. Studies to date have observed the phosphorylation profile of cell cycle regulators at senescence to be consistent with a growth-arrested state. In the course of this chapter, the manner by which cell

DOI: 10.1016/S1566-3124(04)16002-1

cycle regulators influence a cell to undergo cellular replication or cease further proliferation and enter the senescent state will be discussed.

To begin the chapter, a brief overview of the four phases that comprise the cell cycle and a brief description of the complexes that mediate movement from one phase of the cell cycle to the next is provided. Subsequent sections in the chapter attempt to delve more deeply into each of the regulators involved in the cell cycle. The final segment of the chapter attempts to introduce the reader to the phenomenon of cellular senescence and how the activity of cell cycle regulators is altered in the senescent state with a particular focus on protein phosphorylation.

2. Cell cycle overview

The eukaryotic process of cell cycle progression is an exquisitely choreographed molecular performance. Cell division is normally composed of an ordered series of discrete biochemical transitions that coordinate chromosome replication and segregation with cell growth and cytokinesis. The abrupt physiological changes associated with key cell cycle events, such as the initiation of DNA replication or chromosome disjunction, are determined in large part by the timely synthesis or degradation of specific effector proteins. Accurate control of cell division is essential for genomic stability and malignancies are characterized by aberrations in the expression of the molecular mediators that drive or inhibit cellular proliferation (Hanahan and Weinberg, 2000). The actual progression of the cell through the cell cycle involves four different phases: G1, S, G2, and M phase (see Fig. 1). Following mitosis, the daughter cells enter the G1 phase, where cell growth primarily occurs, and it is also during this phase that the cell is responsive to

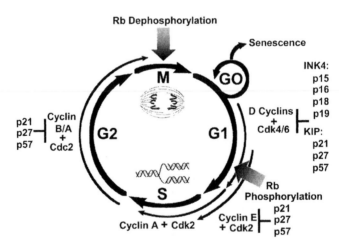

Fig. 1. The cell cycle is comprised of four phases: G1, S (DNA replication), G2, and M (mitosis) phase. Cells can exit the cell cycle and enter a quiescent G0 state upon serum deprivation or cellular senescence after sufficient population doublings. Movement of the cell from one phase to the next is positively regulated by a family of cyclin/cdk complexes and cdk activities are negatively regulated by the KIP and INK families of cdk inhibitors.

growth inhibitory or stimulatory signals. The G1 phase is followed by the S phase, in which DNA replication occurs. This is immediately followed by the G2 phase, during which the integrity of DNA replication is verified. The cells then reenter M phase and undergo mitosis to produce two daughter cells. Cells can exit the cell cycle upon the removal of serum or growth factors and enter a quiescent or G0 state, which is characterized by a low rate of metabolic activity and sensitivity to mitogens. Alternatively, cells at the end of their proliferative life span exit the cell cycle and enter a proliferatively inert state referred to as cellular senescence.

Movement of the cell through the cell cycle requires the activity of a family of cyclin-dependent kinases (cdks) that target serine and threonine residues for phosphorylation. The cdks consist of a family of related kinases, cdks 1–7 whose members require the association with a cyclin for catalytic activation (Sherr, 1994; Morgan, 1995). Different cyclins bind and activate different cdks. The transition from the G1 phase into S phase requires the phosphorylation of the retinoblastoma protein (pRb) which is mediated primarily by cyclin D (cyclins D1, D2, and D3) associated with cdk4 or cdk6, and also by E-type cyclins (cyclins E1 and E2) in association with cdk2 (Dowdy et al., 1993; Ewen et al., 1993). Phosphorylation of pRb releases associated E2F–DP1 heterodimers from inhibition of its transcriptional activity (Pagano et al., 1992a; Kato et al., 1993).

Progression through G1, and entrance and progression through S phase is dependent upon cdk2 kinase activity (Parker et al., 1991; Dulic et al., 1992; Koff et al., 1992; Lees et al., 1992). In late G1 and early S phase, cyclin E–cdk2 and the more recently discovered cyclin E2/cdk2 phosphorylate various targets which are required for S phase entrance (Lauper et al., 1998; Gudas et al., 1999). Cyclin D1 and E-type cyclins are essential for the movement of the cell through the G1/S transition. Microinjection of antibodies to cyclin D1 (Baldin et al., 1993), cyclin E1 (Ohtsubo and Roberts, 1993), or cyclin E2 (Gudas et al., 1999) can inhibit G1 to S phase progression. Overexpression of cyclin D1 or cyclin E shortens the G1/S phase interval (Resnitzky et al., 1994). Moreover, simultaneous coexpression of both cyclins further shortens the G1 phase, suggesting that cyclin D1 and E complexes may have distinct targets (Resnitzky and Reed, 1995). Work in cyclin E → D1 knockin mice, where the cyclin D1 gene has been replaced with cyclin E has demonstrated that cyclin E can replace the activity of cyclin D1 (Geng et al., 1999). In late S phase, cdk2 becomes associated with cyclin A. The activity of the cyclin A–cdk2 complex is essential for entrance to and progression through S phase (Giordano et al., 1989; Pagano et al., 1992b). Movement through the final cell cycle checkpoint at the G2/M transition is mediated by the activity of cdk1 in association with partners of the cyclin B family.

3. pRB- and E2F-mediated transcriptional control

The E2F family of transcription factors activate and repress the expression of various genes to promote cell cycle progression into S phase (Cam and Dynlacht, 2003). A heterodimeric complex consisting of an E2F transcription factor and its dimerization partner, a DP protein, binds to specific E2F DNA-binding sites and leads to a modulation of promoter activity. In early G1, and during G0, the association of pRb with E2F complexes is mediated by the phosphorylation status of pRB, were only a hypophosphorylated form of pRb associates with E2F complexes (Stevaux and Dyson, 2002).

The recruitment of pRb proteins to cdk/cyclin complexes occurs via the interaction between cyclins D, E, and A with pRb, leading to phosphorylation on various serine and threonine residues. This phosphorylation of pRb results in its dissociation from E2F complexes and the subsequent induction of genes that are required for continued cell cycle progression. Viral oncoproteins such as SV40 Large-T Antigen, Human Papillomavirus E7, and Adenoviral E1A have been shown to bind pRb and consequently stimulate quiescent cells to enter a proliferatively active state. Furthermore, mutations of pRb have been frequently identified in various tumor types to evidence its role as a critical tumor suppressor.

4. Cyclins: A requirement for cdk activation

As the name cdk implies, the binding of a cyclin is an absolute requirement for the catalytic activation of the cdks. The cyclins consist of a family of proteins sharing a conserved region of approximately 100 amino acids, referred to as the cyclin box (Sherr, 1994). The cyclin box serves as a docking site for the recruitment of cdk inhibitors or cdk substrates to cyclin/cdk complexes. Cyclin binding leads to conformational changes within the cdk that are required for its activation (Jeffrey et al., 1995). These changes include a reorientation of a glutamate residue within the cdk active site such that ATP binding and recruitment is enhanced. In addition, rearrangement of the cdk T loop occurs to reduce steric interference upon substrate binding (De Bondt et al., 1993; Jeffrey et al., 1995).

The precise timing of cyclin/cdk activation during the cell cycle ensures that the complexes are catalytically active only when required. This is regulated in part by the specific subcellular localization and the timed expression of the various cyclins throughout the cell cycle. In general, peak nuclear localization of a specific cyclin occurs when peak activity of the partner kinase is required.

Cyclin levels are regulated by ubiquitin-mediated proteolysis (Slingerland and Pagano, 1997). The cyclins are rapidly exported from the nucleus and degraded when not required (Willems et al., 1996; Diehl and Sherr, 1998). For example, the cyclin D1 protein accumulates in the nucleus during G1 and exits the nucleus and is degraded during late G1 and S phase (Baldin et al., 1993; Diehl et al., 1997; Diehl and Sherr, 1998). Similarly, nuclear import and export of B-type cyclins are actively regulated in a cell cycle-dependent manner. Phosphorylation of specific sites in cyclin B1 are required for association of nuclear export factor CRM1 (Yang et al., 1998; Moore et al., 1999). Synergistic coupling of nuclear export and degradation of cyclins is triggered by protein phosphorylation to tightly regulate cyclin availability. Phosphorylation at thr-286 on cyclin D1 by GSK3-β functions to trigger its export from the nucleus- and ubiquitin-dependent proteolysis (Diehl et al., 1997; Diehl and Sherr, 1998). Mutant forms of cyclin D, T286A remain localized in the nucleus and had half-lives approaching 3.5 h, as opposed to a half-life of

25 min in asynchronously growing cells (Diehl et al., 1997). It is possible that GSK3-β-mediated phosphorylation of cyclin D1 may be necessary for crm1 association. The degradation of cyclin E is similarly regulated by phosphorylation on a thr-380 residue. Cyclin E/cdk2 complexes phosphorylate the thr-380 residue to target cyclin E for ubiquitin-dependent proteolysis (Clurman et al., 1996). Cyclin A and B degradation is dependent upon a sequence located at the amino terminus, referred to as the destruction box. Removal of this motif functions to stabilize these two cyclins (Glotzer et al., 1991).

Cyclins are also subject to important regulation at the level of transcription. Growth factors have been shown to stimulate cyclin D1 synthesis and their removal reduces its mRNA synthesis (Matsushime et al., 1991; Muller et al., 1994; Cheng et al., 1998). Cyclins E, A, and B are also transcriptionally regulated, with peak mRNA expression coinciding with peak kinase activities (Sherr, 1994).

5. Cdk inhibitors regulate cdk activity

The cdk inhibitors consist of two families: *in*hibitors of cd*k4* (INK4) and the *k*inase *in*hibitor *proteins* (KIPs) families (Sherr and Roberts, 1995). The Kip family consists of three broadly acting inhibitors p21^{CIP1}, p27^{KIP1}, and p57^{KIP2}. Kip family members bind to and inhibit the cyclin–cdk complex (El-Deiry et al., 1993; Harper et al., 1993, 1995). *In vitro* experiments demonstrate that a single molecule of p21 or p27 is sufficient to inhibit the kinase activity of cyclin/cdk complexes (Hengst et al., 1998). The affinity with which these KIPs associate with target cyclin/cdks may be regulated by phosphorylation (Kawada et al., 1997; Sheaff et al., 1997). Inhibition of cyclin D- and cyclin E-associated kinases by KIPs results in a G1 cell cycle arrest. p27 was first identified in cells arrested by transforming growth factor-β (TGF-β), by contact inhibition and by lovastatin (Koff et al., 1993; Hengst et al., 1994; Polyak et al., 1994; Slingerland et al., 1994). p27 is essential for G0 arrest resulting from serum withdrawal (Coats et al., 1996) and upon estradiol withdrawal from estrogen-dependent breast cancer cells (Cariou et al., 2000). An increase in the association of p27^{KIP1} with its target cyclin/cdks occurs in response to a number of antiproliferative signals. Levels of both p27 and p21 are regulated by ubiquitin-mediated proteolysis (Pagano et al., 1995; Maki and Howley, 1997; Tsvetkov et al., 1999). p27 is also subject to translational controls (Hengst and Reed, 1996; Millard et al., 1997). p21 is induced by many forms of cellular stress and by DNA damage, and p21 upregulation serves to coordinate cell cycle arrest with mechanisms of DNA repair (Dulic et al., 1994; Waga et al., 1994). In addition to their role as cdk inhibitors, the KIPs may also facilitate assembly and localization of cyclin D1/cdk complexes within the nucleus (LaBaer et al., 1997; Cheng et al., 1999).

The inability of cyclin/cdk complexes to localize within the nucleus upon their overexpression identifies the need for a nuclear chaperone (LaBaer et al., 1997). The overexpression of p21 or p27 is required to localize ectopically expressed cyclin D/cdk4 into the nucleus to suggest that the KIP inhibitors function to chaperone cyclin/cdk complexes into the nucleus (LaBaer et al., 1997). The presence of a nuclear localization signal on KIP inhibitors allows for these cdk inhibitors to function as nuclear chaperones. The cytosolic localization of the KIP inhibitors is in part mediated by Akt-targeted phosphorylation of p27 to prevent its nuclear localization (Liang et al., 2002).

The formation of cyclin/cdk complexes *in vitro* and *in vivo* has been shown to require the presence of KIP inhibitors. In mouse embryonic fibroblasts lacking either p21, p27, or p21 and p27, the formation of cyclin D/cdk4 complexes was found to be substantially perturbed (Cheng et al., 1999). The reintroduction of the KIP inhibitors was found to restore cyclin D/cdk4 complexes at levels approaching wild-type cells (Cheng et al., 1999). Thus, in addition to their other roles, the KIP inhibitors also function to promote assembly of cyclin/cdk complexes.

The ability to immunoprecipitate p21 and p27 complexes that contain kinase activity indicates that the KIP inhibitors can be bound to cyclin/cdk complexes, and render them in either an active or inactive state (LaBaer et al., 1997). The question that naturally arises from these observations is: what determines whether the KIP inhibitors associate with the complex to activate it or inhibit the complex? *In vitro* assays have demonstrated the ability of p27 to associate with cyclin E–cdk2 complexes in one of two states: with high or low affinity, where the affinity with which p27 associates with target complexes is correlated with its inhibitory activity and a high binding affinity results in cdk inhibition (Sheaff et al., 1997). The affinity of the interaction has been shown to be dependent on the concentration of ATP in the binding assay. The tight binding of p27 to cyclin–cdk complex at low ATP concentrations was correlated with a loss of kinase activity, whereas a lower affinity of p27 binding at elevated ATP concentrations was correlated with p27 functioning as a substrate for the kinase (Sheaff et al., 1997). Alternatively, as yet unidentified, phosphorylation sites on KIP inhibitors may exist to determine the affinity with which they bind to target cyclin/cdk complexes (Ciarallo et al., 2002).

6. INK family of inhibitors

In contrast to the KIP family of inhibitors, the inhibitory activity of INK4 family members, p15^{INK4B}, p16^{INK4A}, p18^{INK4C}, and p19^{INK4D}, is restricted to cdk4 and cdk6 (Sherr and Roberts, 1995; Ruas and Peters, 1998). The p16^{INK4A} gene was discovered to encode a cdk4-associated protein (Serrano et al., 1993) and as the MTS1 gene targeted by chromosomal deletions in many human cancers (Cairns et al., 1994). A tumor-suppressor role for p16 is supported by p16^{INK4A} ARF$^{-/-}$ mice, which have a substantially increased incidence of cancers when compared to wild-type mice (Serrano et al., 1996). p16 levels are increased in senescent fibroblasts and p16 plays a role in the arrest of cellular proliferation at senescence (Alcorta et al., 1996; Hara et al., 1996; Serrano et al., 1996). p15^{INK4B} was cloned as a gene induced in response to TGF-β (Hannon and Beach, 1994). In epithelial cells, G1 arrest by TFG-β involves induction of the p15^{INK4B} gene, stabilization of the p15 protein, and accumulation of p15 in cdk4 and cdk6 complexes (Reynisdottir et al., 1995; Sandhu et al., 1997). p18 and p19 were identified in yeast two-hybrid screens that used cdk4 as bait. p18 has been reported to mediate inhibition of cdk4 and cdk6 complexes during cellular differentiation (Phelps and Xiong, 1998).

7. Cdk phosphorylation

Specific phosphorylation and dephosphorylation events are required for cdk kinase activity (see Fig. 2) (Solomon et al., 1992). A thr-160 residue within the T loop of cdk2 (thr-161 in cdk1, and thr-172 in cdk4) must be phosphorylated for the kinase to be active (Gu et al., 1992; Kato et al., 1994). Phosphorylation of this residue induces conformational changes that reduce steric hindrance of substrate binding (Russo et al., 1996). The cdk-activating kinase, or CAK, catalyzes phosphorylation of the thr-160 site. In mammalian cells, CAK is a complex of cdk7 in association with cyclin H. A relatively low conservation of the cyclin box sequence in cyclin H precludes the binding of KIP inhibitors to this cyclin (Fisher and Morgan, 1994; Kato et al., 1994). Consequently, the KIPs do not directly regulate the activity of cdk7/cyclin H complexes (CAK). However, the KIPs may regulate substrate availability for CAK complexes, since the association of KIPs

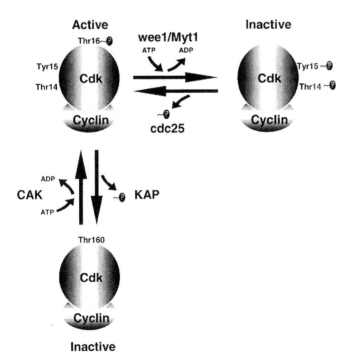

Fig. 2. Regulation of cdk activity by phosphorylation. Phosphorylation of the thr-160 residue (thr-161 in cdk1 and thr-172 in cdk4) and dephosphorylation of the inhibitory sites at thr-14 and tyr-15 (tyr-17 and cdk4) residues are required for cdk activation.

to cyclin/cdks impedes access of CAK to the catalytic cleft of the cdk moiety (Kaldis et al., 1998).

Phosphorylation of thr-14 and/or tyrosine (tyr-15) residues inactivates cdk1 and cdk2, and an analogous inhibitory site exists on cdk4 at tyrosine-17 (Gould and Nurse, 1989; Krek and Nigg, 1991; Gu et al., 1992; Terada et al., 1995). The human homologs of the yeast wee1 and the *Xenopus* myt1 phosphorylate cdc2 at the thr-14 and tyr-15 (Igarashi et al., 1991; Mueller et al., 1995; Booher et al., 1997; Liu et al., 1997).

Members of the cdc25 phosphatase family, cdc25A, cdc25B, and cdc25C, remove the inhibitory phosphates from the cdks (Dunphy and Kumagai, 1991; Galaktinov and Beach, 1991; Gautier et al., 1991; Kumagai and Dunphy, 1991; Strausfeld et al., 1991; Millar and Russell, 1992). These phosphatases are periodically expressed and catalytically active at discrete times within the cell cycle. Cdc25A levels are maximal at the G1/S transition and this phosphatase target cyclin E/cdk2 (Hoffmann et al., 1994; Jinno et al., 1994). Cdc25A may also act on cdk6 (Iavarone and Massague, 1997). Cdc25A expression is induced in certain cell types by c-myc and cdc25A may be phosphorylated and activated by Raf1 or by cyclin E/cdk2 (Hoffmann et al., 1994; Galaktinov et al., 1995, 1996). Cdc25B and cdc25C are predominantly active at the G2/M transition and act on mitotic cyclin B-associated cdk1 (Draetta and Eckstein, 1997).

The cellular localization of cdc25 has been shown to be regulated by the binding of the 14-3-3 family of proteins (Lopez-Girona et al., 1999). In response to radiation-induced DNA damage, chk1 has been shown to phosphorylate cdc25A, cdc25B, and cdc25C at the ser-216 residue. Phosphorylation at ser-216 functions to create a binding site for 14-3-3 proteins (Furnari et al., 1997; Peng et al., 1997; Sanchez et al., 1997). The binding of 14-3-3 proteins to cdc25 is associated with a loss of phosphatase activity, and consequently aids in arresting cellular proliferation. Studies in fission yeast have identified that the binding of Rad24, a member of the 14-3-3 family of proteins, functions to regulate the cellular localization of cdc25 in response to DNA damage (Lopez-Girona et al., 1999). The absence of a nuclear export signal (NES) within cdc25 precludes its transit from the nucleus by direct binding with crm1, a nuclear export factor (Fornerod et al., 1997). Instead, the presence of an NES on Rad24 allows for the recruitment of crm1, which binds specifically to proteins containing NES. Thus, Rad24 functions to chaperone cdc25 out of the nucleus and into the cytoplasm by virtue of its NES. The subsequent loss of cdc25 from the nucleus results in an accumulation of inhibitory phosphorylation on cdks and their subsequent inhibition.

The recruitment of cdc25 phosphatases to cyclin/cdk complexes occurs via an interaction with the cyclin box (Saha et al., 1997). A region at the N-terminus of cdc25 between amino acids 10 and 14 has been shown to contain an amino acid sequence of RRLLF, a sequence that is highly homologous to the RRLFG consensus sequence that confers binding to the cyclin box on cyclins. Consequently, it was postulated that the presence of this site on cdc25A confers the ability to bind to cyclins (Saha et al., 1997). Site-directed mutagenesis of the RRLLF sequence was shown to abolish cyclin binding by cdc25A, and a peptide containing the cyclin box consensus binding sequence was shown to competitively inhibit the association of cdc25A with cyclin E or A (Saha et al., 1997). Therefore, cdk complexes that contain an associated KIP molecule cannot be dephosphorylated by cdc25 due to the inability of the phosphatase to bind to the complex (Saha et al., 1997).

8. Cellular senescence

The proliferative potential of normal cells in culture is constrained to a finite number of population doublings (Hayflick, 1965). The eventual growth arrest that defines the termination of cellular proliferation is referred to as cellular senescence. Upon entering the senescent state, cultured cells remain viable if maintained by the periodic repletion of culture media (Hayflick, 1965; La Thangue, 1994). The point of entry into the senescent state is not determined by the duration of time that cells have existed, but instead by the number of population doublings that the cells have undergone (Hayflick and Moorhead, 1961). Furthermore, entrance into the senescent state is a gradual process in any cultured cell population, where the percentage of cells that are senescent increases steadily with increasing population doublings (Cristofalo and Sharf, 1973). Thus, early passage cells contain a small number of senescent cells and at later passages, the population is progressively dominated by senescent cells.

That senescence is not only an *in vitro* phenomenon, but also occurs *in vivo*, is indicated by several observations. To begin with, the proliferative capacity of finite life-span cells in culture is inversely related to the donor's age, such that with increasing donor's age there is a decreased proliferative capacity of cells in culture (Martin et al., 1970). The number of population doublings that cells will undergo in culture is also related directly to the longevity of the donor species (Rohme, 1981). Cells cultured from long-lived species, such as the Galapagos tortoise, have a greater proliferative capacity than shorter-lived species such as mice (Rohme, 1981). The identification of senescent cells in tissue culture and biopsy samples was aided by the observation that senescent cells acquire the unique property of acid-stable β-galactosidase (Dimri et al., 1995). Thus, the incubation under acidic conditions of senescent cells with the β-galactosidase substrate, X-gal served to stain senescent cells blue at pH 6.0. In biopsies from human donors of different ages, there was an age-dependent increase of β-galactosidase in dermal fibroblasts and in epidermal keratinocytes, indicating an accumulation of senescent cells during the aging process *in vivo* (Dimri et al., 1995).

A model of senescence used to describe the phenomenon in fibroblasts and epithelial cells includes two stages (Wright and Shay, 1992). In the initial stage, cells will undergo approximately 70 rounds of cell division. Thereafter, the cells will enter a proliferatively inert state, senescence that is characterized by cells staining positive for β-galactosidase and maintaining a large flattened morphology. Escape from this initial senescence arrest is mediated by expressing various viral oncogenes (Shay et al., 1991a). Experiments involving the transfection of fibroblasts and epithelial cells with viral oncogenes targeting pRb and p53, such as SV40 Large-T

Antigen, Human Papilloma Virus (HPV16) oncogenes E6 and E7, or Adenovirus type 5 oncogenes E1a and E2a, have resulted in a period of "extended life" (Shay et al., 1991a, 1993; Wright and Shay, 1992; Jarrard et al., 1999). Escape from the initial senescence arrest point has been shown to have tissue-specific requirements. In human mammary epithelial cells (HMECs), overexpression of introduced HPV16 E6, which functions to bind to and inhibit p53 function, can serve to circumvent the arrest point (Shay et al., 1993). However, in human fibroblasts both HPV16 E6 and E7, the latter of which sequesters pRb, are required to circumvent senescence and thus allow fibroblasts to enter a period of extended growth (Shay et al., 1991a).

This period of extended life involves continued proliferation of cells beyond the initial senescence arrest point by approximately 20 population doublings. Thereafter, the cells enter a state of "crisis" where cell death is balanced by cell proliferation to result in no net increase in the number of cells (Wright and Shay, 1992). Infrequently, cells may spontaneously emerge from crisis to form immortal clones (Shay et al., 1991b).

The relevance of senescence to cancer emanates from the suggestion that it may function as a tumor-suppression mechanism. It has been theorized that a normal cell can only form a tumor of 1 cm^3 prior to reaching the proliferative constraints imposed by senescence (Vojta and Barrett, 1995). Thus, in order for tumor cells to evolve into a metastatic state, it would require them to overcome the senescence checkpoint.

The reduction in telomere length that occurs in response to DNA replication is widely believed to be the trigger for the entry of cells into senescence (Harley and Villeponteau, 1995), although telomere-independent mechanisms of senescence have been reported (Mathon and Lloyd, 2001). In humans, telomeres consist of a repeating double-stranded DNA sequence of TTAGGG located at the termini of chromosomes (Moyzis et al., 1988; Allshire et al., 1989; Meyne et al., 1989). The semiconservative nature of DNA replication, involves a leading strand and a lagging strand, which requires numerous RNA primers, results in the formation of an RNA–DNA duplex at the terminus of the lagging strand. The requirement of a RNA primer for DNA polymerase activity precludes the formation of a DNA duplex, and consequently results in the formation of a single-stranded 3'OH tail that is degraded (Harley et al., 1990; Greider, 1998). Thus after each round of DNA replication, in most somatic cells, the telomere sequence decreases by approximately 50–100 basepairs to yield a progressively smaller chromosome (Harley et al., 1990). Ultimately, it is widely believed that the telomeres reach a critical length, which is thought to trigger a series of events leading to G1 cell cycle arrest at senescence.

Telomere length is thought to be regulated by an assortment of telomere-binding proteins and an enzyme, telomerase (Greider and Blackburn, 1985). Telomerase functions to extend the 3'OH single-strand DNA end. Extension of the 3' end provides additional sequence for the annealing of RNA primers and the subsequent DNA polymerization reaction (Greider and Blackburn, 1989). This functions to preserve the telomeres, and consequently allows for the retention of a cell's chromosomal endowment. In addition, the maintenance of telomere lengths serves to prevent the accumulation of dicentric chromosomes, and other chromosomal abnormalities that occur upon a reduction in telomere length (Zhu et al., 1999). The enzyme itself, was found to be a ribonucleoprotein, consisting of an RNA and protein component (Greider and Blackburn, 1989). The RNA component of telomerase consists of a region of sequence homology with the repeating telomere sequence. Association of the RNA template with a single-strand DNA overhang functions to anneal and align the enzyme to telomeres. The protein component of telomerase contains a reverse transcriptase domain that utilizes the RNA component of the enzyme as a template to catalyze the addition of nucleotides to the single-stranded 3′ tail (Harrington et al., 1997; Meyerson et al., 1997; Nakamura et al., 1997).

Prior studies have identified that most normal somatic cells express the RNA component of telomerase but fail to express hTERT, the protein component (Autexier et al., 1996). Consequently, the introduction of hTERT alone into human fibroblasts was sufficient to result in the induction of catalytic telomerase activity, and result in cellular immortalization (Bodnar et al., 1998; Vaziri and Benchimol, 1998). However, studies in other cell types have demonstrated that the presence of hTERT was not sufficient for cellular immortalization. In HMECs, inactivation of the pRb/p16 pathway by either HPV E7 or p16 promoter methylation was required in addition to hTERT for cellular immortalization (Kiyono et al., 1998). Upstream regulators of hTERT expression such as c-myc have similarly been shown to immortalize HMECs by virtue of their ability to induce hTERT expression in HMEC, where the pRB/p16 pathway has been abrogated (Wang et al., 1998; Wu et al., 1999). Although hTERT has been reported to immortalize, it is not sufficient to confer oncogenic transformation (Bodnar et al., 1998). In addition to hTERT, expression of SV40 Large-T antigen and an oncogenic allele of H-Ras were required for the oncogenic transformation of human epithelial and fibroblast cells (Hahn et al., 1999).

The manner by which the expression of the cell cycle inhibitors are induced upon telomere shortening remains unresolved. The identification of the molecules that mediate this telomere-induced signal would serve to identify an interface between the cell cycle and the chromosomal status of a cell. To date, it has been proposed that a shortening in telomere length may serve to sufficiently alter the chromosomal confirmation such that previously silent genes may now be expressed (Shay, 1995).

9. Cell cycle, senescence, and protein phosphorylation

Protein phosphorylation is a key mechanism by which intracellular signaling is mediated. In cell cycle progression, phosphoester-containing cell cycle regulators are created by phosphorylation of serine/threonine or tyrosine residues. The resultant phosphoforms undergo cofirmational changes to affect their biochemical activity. In the course of this chapter, cdk targets for protein phosphorylation and the manner by which cell cycle regulators themselves are regulated by phosphorylation has been explored. Here, we will now discuss the relationship between protein phosphorylation, cell cycle, and senescence. To date, only a limited number of cell cycle proteins have been identified as being altered in their phosphorylation status at senescence. This is in part attributable to the technical limitations of phosphate labeling in senescent cells which poorly incorporate phosphate label into proteins due to the relatively low level of metabolic activity at senescence.

Evidence that protein phosphorylation plays a role in cellular senescence was first established by the observation that pRb exists in a hypophosphorylated state in senescent human fibroblasts (Stein et al., 1990). In this early study examining pRb, phosphorylation status was assessed by virtue of mobility shifts on SDS-PAGE and treatment with alkaline phosphatase. Further, investigations of pRb lead to the identification of additional sites whose phosphorylation failed to manifest as mobility shifts on SDS-PAGE (Ezhevsky et al., 1997). One such site was at pRb–ser-780, an exclusive target of cyclin D1/cdk4 complexes in in vitro assays. Stein et al. (1999) reported that pRb recovered from senescent human fibroblasts was not recognized by a ser-780 phospho-specific antibody. These findings were consistent with an absence of cdk activity at senescence.

From this point, several studies over the past decade have attempted to identify the mechanism of cdk inactivation at senescence. For the most part, these studies attributed cdk inactivation to an accumulation of cdk inhibitors at senescence with the proposed senescent inhibitors being a function of the cell type (Alcorta et al., 1996; Hara et al., 1996; Reznikoff et al., 1996; Palmero et al., 1997; Zindy et al., 1997; Collins and Sedivy, 2003). An increased expression of p21 and/or p16 at senescence has been identified in human and murine fibroblasts and melanocytes. The induction of p21 in senescent fibroblasts led to the initial identification and cloning of this gene (Noda et al., 1994). The elevated expression of p16 and/or p21 at senescence is associated with their increased binding and inhibition of G1/S phase-associated cdks. An elimination of p21 expression through homologous recombination extended the life span of human diploid fibroblasts in culture (Brown et al., 1997). Hence, loss of p21 expression, while itself not sufficient to abrogate senescence arrest, may represent a key step in immortalization of cells of fibroblastic lineage. The frequent loss of p21 and p16 expression in human cancers suggests that in vivo, these inhibitors may contribute to the senescence checkpoint and limit tumor development (Cairns et al., 1995).

To date, a single study has examined the phosphorylation status of cdk inhibitors at senescence. In a tissue culture model of prostatic epithelial cells, an increase in p16 levels and binding to target cdks at senescence was correlated with changes in its phosphorylation profile (Sandhu et al., 2000b). Currently, the exact sites targeted for phosphorylation in p16 are unknown. Presumably, phosphorylation of p16 functions to regulate p16 cellular localization, proteolysis, and potentially cdk-binding activity. Sequence- and structure-based analysis of p16 leads to four predicted phosphorylation sites: three serine sites (S7, S8, and S43) and one threonine residue (T93) (Blom et al., 1999). Future mutational analysis will verify the functionally significant sites for p16 phosphorylation and their role in senescence.

A lack of cyclin expression does not appear to be involved in the inhibition of cdk complexes at senescence, since it has been reported that cyclin D and E steady-state levels were substantially elevated in senescent fibroblasts (Dulic et al., 1993; Lucibello et al., 1993; Robetorye et al., 1996). However, the expression of cyclin A and cdk2 are greatly reduced in senescent fibroblasts from both human and rodent models (Afshari et al., 1993). This loss of cyclin A at senescence is likely a consequence of the cells arresting at a point within the cell cycle prior to cyclin A induction.

Investigations of cdk inactivation at senescence have examined the phosphorylation status of cdks (Dulic et al., 1993; Sandhu et al., 2000a). To date, these studies have focused exclusively on cdk2. This is in part attributable to the relative ease by which the phosphorylation status of cdk2 can be assessed by readily available immunoblotting agents. Reports to date have observed the total level of cdk phosphorylation at thr-160 to be decreased at senescence (Stein et al., 1999; Sandhu et al., 2000a). However, the fraction of cdk2 bound to cyclin E was observed to be unaltered at senescence (Sandhu et al., 2000a). Thus, a lack of phosphorylation at this site is unlikely to be the cause of cdk inhibition in senescent cells. Alternatively, phosphorylation at cdk inhibitory sites may function to effect cdk activity at senescence. The evidence in support of this notion includes a study using a HMEC model (Sandhu et al., 2000a). In this study, cyclin E–cdk2 complexes were recovered from senescent cells by immunoprecipitation. These complexes were subsequently treated with recombinant cdc25A phosphatase which partially restored cdk kinase activity. The absence of phosphatase activity in these cells was attributed to a lack of cdc25A expression observed on both western and northern analysis of extract derived from senescent cells.

10. Summary and future directions

Protein phosphorylation plays a central role in the regulation of biochemical processes within the cell. In the context of cell cycle regulation, cell growth and division is permitted only by the appropriate sequence of protein phosphorylation. Initial studies of senescent cells have observed changes in protein phosphorylation that contribute toward cdk inactivation and presumably cell cycle arrest at senescence. The evidence for phosphorylation in mediating senescence stems from two points. To begin with, cdk complexes at senescence maintain a phosphorylation profile that is incompatible with cdk activity. Secondly, cdk inhibitors are regulated at the level of proteolysis and cellular localization by protein phosphorylation. Hence, both the accumulation of cdk inhibitors and their increased binding to cdk complexes is consistent with alterations in their phosphorylation. This suggests the need for a specific phosphorylation profile for senescent proteins to mediate their increased steady-state level and binding to cdk complexes.

Currently, the identification of sites targeted for protein phosphorylation remains an arduous process. Available high throughput methods aid only in the identification of proteins. Additional advances may permit both the identification of proteins and their sites of phosphorylation. An understanding of protein phosphorylation at a proteomic level will undoubtedly lead to a greater understanding of the regulation of cellular processes such as senescence.

References

Afshari, C.A., Vojta, P.J., Annab, L.A., Futreal, P.A., Willard, T.B., Barrett, J.C., 1993. Investigation of the role of G1/S cell cycle mediators in cellular senescence. Exp. Cell Res. 209, 231–237.

Alcorta, D.A., Xiong, Y., Phelps, D., Hannon, G., Beach, D., Barrett, J.C., 1996. Involvement of the cyclin-dependent kinase inhibitor p16 (INK4a) in replicative senescence of normal human fibroblasts. Proc. Natl. Acad. Sci. USA 93, 13742–13747.

Allshire, R.C., Dempster, M., Hastie, N.D., 1989. Human telomeres contain at least three types of G-rich repeat distributed non-randomly. Nucleic Acids Res. 17, 4611–4627.

Autexier, C., Pruzan, R., Funk, W.D., Greider, C.W., 1996. Reconstitution of human telomerase activity and identification of a minimal functional region of the human telomerase RNA. EMBO J. 15, 5928–5935.

Baldin, V., Lukas, J., Marcote, M.J., Pagano, M., Draetta, G., 1993. Cyclin D1 is a nuclear protein required for cell cycle progression in G1. Genes Dev. 7, 812–821.

Blom, N., Gammeltoft, S., Brunak, S., 1999. Sequence and structure-based prediction of eukaryotic protein phosphorylation sites. J. Mol. Biol. 294, 1351–1362.

Bodnar, A.G., Ouellette, M., Frolkis, M., Holt, S.E., Chiu, C.P., Morin, G.B., Harley, C.B., Shay, J.W., Lichtsteiner, S., Wright, W.E., 1998. Extension of life-span by introduction of telomerase into normal human cells. Science 279, 349–352.

Booher, R.N., Holman, P.S., Fattaey, A., 1997. Human Myt1 is a cell cycle-regulated kinase that inhibits Cdc2 but not Cdk2 activity. J. Biol. Chem. 272, 22300–22306.

Brown, J.P., Wei, W., Sedivy, J.M., 1997. Bypass of senescence after disruption of $p21^{CIP1/WAF1}$ gene in normal diploid human fibroblasts. Science 277, 831–834.

Cairns, P., Mao, L., Merlo, A., Lee, D., Schwab, D., Eby, Y., Tokino, K., van der Riet, P., Blaugrund, J.E., Sidransky, D., 1994. Rates of p16 (MTS1) mutations in primary tumors with 9p loss. Science 265, 415–417.

Cairns, P., Polascik, T.J., Eby, Y., Tokino, K., Califano, J., Merlo, A., Mao, L., Herath, J., Jenkins, R., Westra, W., Rutter, J.L., Buckler, A., Gabrielson, E., Tockman, M., Cho, K.R., Hedrick, L., Bova, G.S., Isaacs, W., Koch, W., Schwab, D., Sidransky, D., 1995. Frequency of homozygous deletion at p16/CDKN2 in primary human tumours. Nat. Genet. 11, 210–212.

Cam, H., Dynlacht, B.D., 2003 Apr. Emerging roles for E2F: beyond the G1/S transition and DNA replication. Cancer Cell 3(4), 311–316.

Cariou, S., Donovan, J.C., Flanagan, W.M., Milic, A., Bhattacharya, N., Slingerland, J.M., 2000. Down-regulation of p21WAF1/CIP1 or p27Kip1 abrogates antiestrogen-mediated cell cycle arrest in human breast cancer cells. Proc. Natl. Acad. Sci. USA 97(16), 9042–9046.

Cheng, M., Sexl, V., Sherr, C.J., Roussel, M.F., 1998. Assembly of cyclin D-dependent kinase and titration of p27Kip1 regulated by mitogen-activated protein kinase kinase (MEK1). Proc. Natl. Acad. Sci. USA 95, 1091–1096.

Cheng, M., Olivier, P., Diehl, J.A., Fero, M., Roussel, M.F., Roberts, J.M., Sherr, C.J., 1999. The p21(Cip1) and p27(Kip1) CDK "inhibitors" are essential activators of cyclin D-dependent kinases in murine fibroblasts. EMBO J. 18, 1571–1583.

Ciarallo, S., Subramaniam, V., Hung, W., Lee, J.H., Kotchetkov, R., Sandhu, C., Milic, A., Slingerland, J.M., 2002. Altered p27(Kip1) phosphorylation, localization, and function in human epithelial cells resistant to transforming growth factor beta-mediated G(1) arrest. Mol. Cell. Biol. 22, 2993–3002.

Clurman, B.E., Sheaff, R.J., Thress, K., Groudine, M., Roberts, J.M., 1996. Turnover of cyclin E by the ubiquitin proteasome pathway is regulated by cdk2 binding and cyclin phosphorylation. Genes Dev. 10, 1979–1990.

Coats, S., Flanagan, M., Nourse, J., Roberts, J.M., 1996. Requirement of p27Kip1 for restriction point control of the fibroblast cell cycle. Science 272, 877–880.

Collins, C.J., Sedivy, J.M., 2003. Involvement of the INK4a/Arf gene locus in senescence. Aging Cell 2, 145–150.

Cristofalo, V.J., Sharf, B.B., 1973. Cellular senescence and DNA synthesis. Thymidine incorporation as a measure of population age in human diploid cells. Exp. Cell Res. 76, 419–427.

De Bondt, H.L., Rosenblatt, J., Jancarik, J., Jones, H.D., Morgan, D.O., Kim, S.H., 1993. Crystal structure of cyclin-dependent kinase 2. Nature 363, 595–602.

Diehl, J.A., Sherr, C.J., 1998. Glycogen synthase kinase-3beta regulates cyclin D1 proteolysis and subcellular localization. Genes Dev. 15, 12(22), 3499–3511.

Diehl, J.A., Zindy, F., Sherr, C.J., 1997. Inhibition of cyclin D1 phosphorylation on threonine-286 prevents its rapid degradation via the ubiquitin-proteasome pathway. Genes Dev. 11, 957–972.

Dimri, G.P., Lee, X., Basile, G., Acosta, M., Scott, G., Roskelley, C., Medrano, E.E., Linskens, M., Rubelj, I., Pereira-Smith, O., 1995. A biomarker that identifies senescent human cells in culture and in aging skin in vivo. Proc. Natl. Acad. Sci. USA 92, 9363–9367.

Dowdy, S.F., Hinds, P.W., Louie, K., Reed, S.I., Arnold, A., Weinberg, R.A., 1993. Physical interaction of the retinoblastoma protein with human D cyclins. Cell 73, 499–511.

Draetta, G., Eckstein, J., 1997. Cdc25 protein phosphatases in cell proliferation. Biochim. Biophys. Acta 1332, M53–M63.

Dulic, V., Lees, E., Reed, S.I., 1992. Association of human cyclin E with a periodic G1-S phase protein kinase. Science 257, 1958–1961.

Dulic, V., Drullinger, L.F., Lees, E., Reed, S.I., Stein, G.H., 1993. Altered regulation of G1 cyclins in senescent human diploid fibroblasts: accumulation of inactive cyclin E/Cdk2 and cyclin D1/Cdk2 complexes. Proc. Natl. Acad. Sci. USA 90, 11034–11038.

Dulic, V., Kaufman, W.K., Wilson, S.J., Tosty, T.D., Lees, E., Harper, J.W., Elledge, S.J., Reed, S.I., 1994. p53-dependent inhibition of cyclin-dependent kinase activities in human fibroblasts during radiation-induced G1 arrest. Cell 76, 1013–1023.

Dunphy, W.G., Kumagai, A., 1991. The cdc25 protein contains an intrinsic phosphatase activity. Cell 67, 189–196.

El-Deiry, W.S., Tokino, T., Velculescu, V.E., Levy, D., Parsons, R., Trent, J.M., Lin, D., Mercer, W.E., Kinzler, K.W., Vogelstein, B., 1993. WAF1, a potential mediator of p53 tumor suppression. Cell 75, 817–825.

Ewen, M.E., Slus, H.K., Sherr, C.J., Matsushime, H., Kato, J.-Y., Livingston, D.M., 1993. Functional interactions of the retinoblastoma protein with mammalian D-type cyclins. Cell 73, 487–497.

Ezhevsky, S.A., Nagahara, H., Vocero-Akbani, A.M., Gius, D.R., Wei, M.C., Dowdy, S.F., 1997. Hypo-phosphorylation of the retinoblastoma protein (pRb) by cyclin D:Cdk4/6 complexes results in active pRb. Proc. Natl. Acad. Sci. USA 94, 10699–10704.

Fisher, R.P., Morgan, D.O., 1994. A novel cyclin associates with MO15/CDK7 to form the CDK-activating kinase. Cell 78, 713–724.

Fornerod, M., Ohno, M., Yoshida, M., Mattaj, I.W., 1997. CRM1 is an export receptor for leucine-rich nuclear export signals. Cell 90, 1051–1060.

Furnari, B., Rhind, N., Russell, P., 1997. Cdc25 mitotic inducer targeted by chk1 DNA damage checkpoint kinase. Science 277, 1495–1497.

Galaktionov, K., Beach, D., 1991. Specific activation of cdc25 tyrosine phosphatases by B-type cyclins: evidence for multiple roles of mitotic cyclins. Cell 67, 1181–1194.

Galaktionov, K., Jessus, C., Beach, D., 1995. Raf1 interaction with Cdc25 phosphatase ties mitogenic signal transduction to cell cycle activation. Genes Dev. 9, 1046–1058.

Galaktionov, K., Chen, X., Beach, D., 1996. Cdc25 cell-cycle phosphatase as a target of c-myc. Nature 382, 511–517.

Gautier, J., Solomon, M.J., Booher, R.N., Bazan, J.F., Kirschner, M.W., 1991. Cdc25 is a specific tyrosine phosphatase that directly activates p34cdc2. Cell 67, 197–211.

Geng, Y., Whoriskey, W., Park, M.Y., Bronson, R.T., Medema, R.H., Li, T., Weinberg, R.A., Sicinski, P., 1999. Rescue of cyclin D1 deficiency by knockin cyclin E. Cell 97, 767–777.

Giordano, A., Whyte, P., Harlow, E., Franza, B.R., Jr., Beach, D., Draetta, G., 1989. A 60 kd cdc2-associated polypeptide complexes with the E1A proteins in adenovirus-infected cells. Cell 58, 981–990.

Glotzer, M., Murray, A., Kirschner, M., 1991. Cyclin is degraded by the ubiquitin pathway. Nature 349, 132–138.

Gould, K., Nurse, P., 1989. Tyrosine phosphorylation of the fission yeast cdc2 + protein kinase regulates entry into mitosis. Nature 342, 39–45.

Greider, C.W., 1998. Telomeres and senescence: the history, the experiment, the future. Curr. Biol. 8, R178–R181.

Greider, C.W., Blackburn, E.H., 1985. Identification of a specific telomere terminal transferase activity in Tetrahymena extracts. Cell 43, 405–413.

Greider, C.W., Blackburn, E.H., 1989. A telomeric sequence in the RNA of Tetrahymena telomerase required for telomere repeat synthesis. Nature 337, 331–337.

Gu, Y., Rosenblatt, J., Morgan, D.O., 1992. Cell cycle regulation of CDK2 activity by phosphorylation of Thr160 and Tyr15. EMBO J. 11, 3995–4005.

Gudas, J.M., Payton, M., Thukral, S., Chen, E., Bass, M., Robinson, M.O., Coats, S., 1999. Cyclin E2, a novel G1 cyclin that binds Cdk2 and is aberrantly expressed in human cancers. Mol. Cell. Biol. 19, 612–622.

Hahn, W.C., Counter, C.M., Lundberg, A.S., Beijersbergen, R.L., Brooks, M.W., Weinberg, R.A., 1999. Creation of human tumour cells with defined genetic elements. Nature 400, 464–468.

Hanahan, D., Weinberg, R.A., 2000. The hallmarks of cancer. Cell 100(1), 57–70.

Hannon, G.J., Beach, D., 1994. p15INK4B is a potential effector of TGF-β-induced cell cycle arrest. Nature 371, 257–261.

Hara, E., Smith, R., Parry, D., Tahara, H., Stone, S., Paters, G., 1996. Regulation of p16CDKN2 expression and its implications for cell immortalization and senescence. Mol. Cell. Biol. 16, 859–867.

Harley, C.B., Villeponteau, B., 1995. Telomeres and telomerase in aging and cancer. Curr. Opin. Genet. Dev. 5, 249–255.

Harley, C.B., Futcher, A.B., Greider, C.W., 1990. Telomeres shorten during ageing of human fibroblasts. Nature 345, 458–460.

Harper, J.W., Adami, G.R., Wei, N., Keyomarsi, K., Elledge, S.J., 1993. The p21 Cdk-interacting protein Cip1 is a potent inhibitor of G1 cyclin-dependant kinases. Cell 75, 805–816.

Harper, J.W., Elledge, S., Keyomarsi, K., Dynlacht, B., Tsai, L.-H., Zhang, P., Dobrowolski, S., Bai, C., Connell-Crowley, L., Swindell, E., Fox, M.P., Wei, N., 1995. Inhibition of cyclin-dependent kinases by p21. Mol. Biol. Cell 6, 387–400.

Harrington, L., Zhou, W., McPhail, T., Oulton, R., Yeung, D.S., Mar, V., Bass, M.B., Robinson, M.O., 1997. Human telomerase contains evolutionarily conserved catalytic and structural subunits. Genes Dev. 11, 3109–3115.

Hayflick, L., 1965. The limited in vitro lifetime of diploid cell strains. Exp. Cell Res. 37, 614–636.

Hayflick, L., Moorhead, P.S., 1961. The serial cultivation of human diploid cell strains. Exp. Cell Res. 25, 585–621.

Hengst, L., Reed, S.I., 1996. Translational control of p27Kip1 accumulation during the cell cycle. Science 271, 1861–1864.

Hengst, L., Dulic, V., Slingerland, J., Lees, E., Reed, S.I., 1994. A cell cycle regulated inhibitor of cyclin dependant kinases. Proc. Natl. Acad. Sci. USA 91, 5291–5294.

Hengst, L., Gopfert, U., Lashuel, H.A., Reed, S.I., 1998. Complete inhibition of Cdk/cyclin by one molecule of p21 (cip 1). Genes Dev. 12(24), 3882–3888.

Hoffmann, I., Draetta, G., Karsenti, E., 1994. Activation of the phosphatase activity of human cdc25A by a cdk2-cyclin E dependent phosphorylation at the G1/S transition. EMBO J. 13, 4302–4310.

Iavarone, A., Massague, J., 1997. Repression of the CDK activator Cdc25A and cell-cycle arrest by cytokine TGF-beta in cells lacking the CDK inhibitor p15. Nature 387, 417–422.

Igarashi, M., Nagata, A., Jinno, S., Suto, K., Okayama, H., 1991. Wee1(+)-like gene in human cells. Nature 353, 80–83.

Jarrard, D.F., Sarkar, S., Shi, Y., Yeager, T.R., Magrane, G., Kinoshita, H., Nassif, N., Meisner, L., Newton, M.A., Waldman, F.M., Reznikoff, C.A., 1999. p16/pRb pathway alterations are required for bypassing senescence in human prostate epithelial cells. Cancer Res. 59, 2957–2964.

Jeffrey, P.D., Russo, A.A., Polyak, K., Gibbs, E., Hurwitz, J., Massague, J., Pavletich, N.P., 1995. Mechanism of CDK activation revealed by the structure of a cyclinA-CDK2 complex. Nature 376, 313–320.

Jinno, S., Suto, K., Nagata, A., Igarashi, M., Kanaoka, Y., Nojima, H., Okayama, H., 1994. Cdc25A is a novel phosphatase functioning early in the cell cycle. EMBO J. 13, 1549–1556.

Kaldis, P., Russo, A.A., Chou, H.S., Pavletich, N.P., Solomon, M.J., 1998. Human and yeast cdk-activating kinases (CAKs) display distinct substrate specificities. Mol. Biol. Cell 9, 2545–2560.

Kato, J.Y., Matsuoka, M., Strom, D.K., Sherr, C.J., 1994. Regulation of cyclin D-dependent kinase 4 (cdk4) by cdk4-activating kinase. Mol. Cell. Biol. 14, 2713–2721.

Kawada, M., Yamagoe, S., Murakami, Y., Suzuki, K., Mizuno, S., Uehara, Y., 1997. Induction of p27Kip1 degradation and anchorage independence by Ras through the MAP kinase signaling pathway. Oncogene 15, 629–637.

Kiyono, T., Foster, S.A., Koop, J.I., McDougall, J.K., Galloway, D.A., Klingelhutz, A.J., 1998. Both Rb/p16INK4a inactivation and telomerase activity are required to immortalize human epithelial cells. Nature 396, 84–88.

Koff, A., Giordano, A., Desai, D., Yamashita, K., Harper, J.W., Elledge, S., Nishimoto, T., Morgan, D.O., Franza, R.B., Roberts, J.M., 1992. Formation and activation of a cyclin E-cdk2 complex during the G1 phase of the human cell cycle. Science 257, 1689–1694.

Koff, A., Ohtsuki, M., Polyak, K., Roberts, J.M., Massague, J., 1993. Negative regulation of G1 in mammalian cells: inhibition of cyclin E-dependent kinase by TGF-β. Science 260, 536–539.

Krek, W., Nigg, E., 1991. Differential phosphorylation of vertebrate p34cdc2 kinase at the G1/S and G2/M transitions of the cell cycle: identification of major phosphorylation sites. EMBO J. 10, 305–316.

Kumagai, A., Dunphy, W.G., 1991. The cdc25 protein controls tyrosine dephosphorylation of the cdc2 protein in a cell-free system. Cell 64, 903–914.

LaBaer, J., Garrett, M.D., Stevenson, L.F., Slingerland, J.M., Sandhu, C., Chou, H.S., Fattaey, A., Harlow, E., 1997. New functional activities for the p21 family of CDK inhibitors. Genes Dev. 11, 847–862.

Lauper, N., Beck, A.R., Cariou, S., Richman, L., Hofmann, K., Reith, W., Slingerland, J.M., Amati, B., 1998. Cyclin E2: a novel CDK2 partner in the late G1 and S phases of the mammalian cell cycle. Oncogene 17, 2637–2643.

Lees, E., Faha, B., Dulic, V., Reed, S.I., Harlow, E., 1992. Cyclin E/cdk2 and cyclin A/cdk2 kinases associate with p107 and E2F in a temporally distinct maner. Genes Dev. 6, 1874–1885.

Liang, J., Zubovitz, J., Petrocelli, T., Kotchetkov, R., Connor, M.K., Han, K., Lee, J.H., Ciarallo, S., Catzavelos, C., Beniston, R., Franssen, E., Slingerland, J.M., 2002. PKB/Akt phosphorylates p27, impairs nuclear import of p27 and opposes p27-mediated G1 arrest. Nat. Med. 8, 1153–1160.

Liu, F., Stanton, J.J., Wu, Z., Piwnica-Worms, H., 1997. The human Myt1 kinase preferentially phosphorylates Cdc2 on threonine 14 and localizes to the endoplasmic reticulum and Golgi complex. Mol. Cell. Biol. 17, 571–583.

Lopez-Girona, A., Furnari, B., Mondesert, O., Russell, P., 1999. Nuclear localization of Cdc25 is regulated by DNA damage and a 14-3-3 protein. Nature 397, 172–175.

Lucibello, F.C., Sewing, A., Brüsselbach, S., Bürger, C., Müller, R., 1993. Deregulation of cyclins D1 and E and suppression of cdk2 and cdk4 in senescent human fibroblasts. J. Cell Sci. 105, 123–133.

Maki, C.G., Howley, P.M., 1997. Ubiquitination of p53 and p21 is differentially affected by ionizing and UV radiation. Mol. Cell. Biol. 17, 355–363.

Martin, G.M., Sprague, C.A., Epstein, C.J., 1970. Replicative life-span of cultivated human cells. Effects of donor's age, tissue, and genotype. Lab. Invest. 23, 86–92.

Mathon, N.F., Lloyd, A.C., 2001. Cell senescence and cancer. Nat. Rev. Cancer 1, 203–213.

Matsushime, H., Roussel, M.F., Ashmun, R.A., Sherr, C.J., 1991. Colony-Stimulating Factor 1 regulates novel cyclins during the G1 phase of the cell cycle. Cell 65, 701–713.

Meyerson, M., Counter, C.M., Easton, E.N., Ellisen, L.W., Steiner, P., Dickinson-Caddle, S., Ziaugra, L., Beijersbergen, R.L., Davidoff, M.J., Liu, Q., Bacchetti, S., Haber, D.A., Weinberg, R.A., 1997. hEST2, the putative human telomerase catalytic subunit gene, is up-regulated in tumor cells and during immortalization. Cell 90, 785–795.

Meyne, J., Ratliff, R.L., Moyzis, R.K., 1989. Conservation of the human telomere sequence (TTAGGG)n among vertebrates. Proc. Natl. Acad. Sci. USA 86, 7049–7053.

Millar, J.B.A., Russell, P., 1992. The cdc25 M-phase inducer: an unconventional protein phosphatase. Cell 68, 407–410.

Millard, S.S., Yan, J.S., Nguyen, H., Pagano, M., Kiyokawa, H., Koff, A., 1997. Enhanced ribosomal association of p27(Kip1) mRNA is a mechanism contributing to accumulation during growth arrest. J. Biol. Chem. 272, 7093–7098.

Moore, J.D., Yang, J., Truant, R., Kornbluth, S., 1999. Nuclear import of Cdk/cyclin complexes: identification of distinct mechanisms for import of Cdk2/cyclin E and Cdc2/cyclin B1. J. Cell Biol. 144, 213–224.

Morgan, D.O., 1995. Principles of Cdk regulation. Nature 374, 131–134.

Moyzis, R.K., Buckingham, J.M., Cram, L.S., Dani, M., Deaven, L.L., Jones, M.D., Meyne, J., Ratliff, R.L., Wu, J.R., 1988. A highly conserved repetitive DNA sequence, (TTAGGG)n, present at the telomeres of human chromosomes. Proc. Natl. Acad. Sci. USA 85, 6622–6626.

Mueller, P.R., Coleman, T.R., Kumagai, A., Dunphy, W.G., 1995. Myt1: a membrane-associated inhibitory kinase that phosphorylates Cdc2 on both threonine-14 and tyrosine-15. Science 270, 86–90.

Muller, H., Lukas, J., Schneider, A., Warthoe, P., Bartek, J., Eilers, M., Srauss, M., 1994. Cyclin D1 expression is regulated by the retinoblastoma protein. Proc. Natl. Acad. Sci. USA 91, 2945–2949.

Nakamura, T.M., Morin, G.B., Chapman, K.B., Weinrich, S.L., Andrews, W.H., Lingner, J., Harley, C.B., Cech, T.R., 1997. Telomerase catalytic subunit homologs from fission yeast and human. Science 277, 955–959.

Noda, A., Ning, Y., Venable, S.F., Pereira-Smith, O.M., Smith, J.R., 1994. Cloning of senescent cell-derived inhibitors of DNA synthesis using an expression screen. Exp. Cell Res. 211, 90–98.

Ohtsubo, M., Roberts, J.M., 1993. Cyclin-dependent regulation of G1 in mammalian cells. Science 259, 1908–1912.

Pagano, M., Draetta, G., Jansen-Dürr, P., 1992a. Association of cdk2 kinase with the transcription factor E2F during S phase. Science 255, 1144–1147.

Pagano, M., Pepperkok, R., Verde, F., Ansorge, W., Draetta, G., 1992b. Cyclin A is required at two points in the human cell cycle. EMBO J. 11, 961–971.

Pagano, M., Tam, S.W., Theodoras, A.M., Beer-Romero, P., Del Sal, G., Chau, V., Yew, P.R., Draetta, G.F., Rolfe, M., 1995. Role of ubiquitin-proteasome pathway in regulating abundance of the cyclin-dependent kinase inhibitor p27. Science 269, 682–685.

Palmero, I., McConnell, B., Parry, D., Brookes, S., Hara, E., Bates, S., Jat, P., Peters, G., 1997. Accumulation of p16INK4a in mouse fibroblasts as a function of replicative senescence and not of retinoblastoma gene status. Oncogene 15, 495–503.

Parker, L.L., Atherton-Fessler, S., Lee, M.S., Ogg, S., Falk, J.L., Swenson, K.I., Piwnica-Worms, H., 1991. Cyclin promotes the tyrosine phosphorylation of p34cdc2 in a weel+ dependent manner. EMBO J. 10, 1255–1263.

Peng, C.Y., Graves, P.R., Thoma, R.S., Wu, Z., Shaw, A.S., Piwnica-Worms, H., 1997. Mitotic and G2 checkpoint control: regulation of 14-3-3 protein binding by phosphorylation of Cdc25C on serine-216. Science 277, 1501–1505.

Phelps, D.E., Xiong, Y., 1998. Regulation of cyclin-dependent kinase 4 during adipogenesis involves switching of cyclin D subunits and concurrent binding of p18INK4c and p27Kip1. Cell Growth Differ. 9, 595–610.

Polyak, K., Kato, J.Y., Solomon, M.J., Sherr, C.J., Massague, J., Roberts, J.M., Koff, A., 1994. p27Kip1, a cyclin-Cdk inhibitor, links transforming growth factor-β and contact inhibition to cell cycle arrest. Genes Dev. 8, 9–22.

Resnitzky, D., Reed, S.I., 1995. Different roles for cyclins D1 and E in regulation of the G1-to-S transition. Mol. Cell. Biol. 3463–3469.

Resnitzky, D., Gossen, M., Bujard, H., Reed, S., 1994. Differential acceleration of the G1/S transition by conditional overexpression of cyclins D1 and E. Mol. Cell. Biol. 14, 1669–1679.

Reynisdottir, I., Polyak, K., Iavarone, A., Massague, J., 1995. Kip/Cip and Ink4 Cdk inhibitors cooperate to induce cell cycle arrest in response to TGF-β. Genes Dev. 9, 1831–1845.

Reznikoff, C.A., Yeager, T.R., Belair, C.D., Savelieva, E., Puthenveettil, J.A., Stadler, W.M., 1996. Elevated p16 at senescence and loss of p16 at immortalization in human papillomavirus 16 E6, but not E7, transformed human uroepithelial cells. Cancer Res. 56, 2886–2890.

Robetorye, R.S., Nakanishi, M., Venable, S.F., Pereira-Smith, O.M., Smith, J.R., 1996. Regulation of p21[Sdi1/Cip1/Waf1/Imda-6] and expression of other cyclin-dependent kinase inhibitors in senescent human cells. Mol. Cell. Differ. 4, 113–126.

Rohme, D., 1981. Evidence for a relationship between longevity of mammalian species and life spans of normal fibroblasts in vitro and erythrocytes in vivo. Proc. Natl. Acad. Sci. USA 78, 5009–5013.

Ruas, M., Peters, G., 1998. The p16INK4a/CDKN2A tumor suppressor and its relatives. Biochim. Biophys. Acta 1378, F115–F177.

Russo, A.A., Jeffrey, P.D., Pavletich, N.P., 1996. Structural basis of cyclin-dependent kinase activation by phosphorylation. Nat. Struct. Biol. 3, 696–700.

Saha, P., Eichbaum, Q., Silberman, E.D., Mayer, B.J., Dutta, A., 1997. p21CIP1 and Cdc25A: competition between an inhibitor and an activator of cyclin-dependent kinases. Mol. Cell. Biol. 17, 4338–4345.

Sanchez, Y., Wong, C., Thoma, R.S., Richman, R., Wu, Z., Piwnica-Worms, H., Elledge, S.J., 1997. Conservation of the Chk1 checkpoint pathway in mammals: linkage of DNA damage to Cdk regulation through Cdc25. Science 277, 1497–1501.

Sandhu, C., Garbe, J., Daksis, J., Pan, C.-H., Bhattacharya, N., Yaswen, P., Koh, J., Slingerland, J., Stampfer, M.R., 1997. Transforming Growth Factor β stabilizes p15[INK4B] protein, increases p15[INK4B]-cdk4 complexes and inhibits cyclin D1/cdk4 association in human mammary epithelial cells. Mol. Cell. Biol. 17, 2458–2467.

Sandhu, C., Donovan, J., Bhattacharya, N., Stampfer, M., Worland, P., Slingerland, J., 2000a. Reduction of Cdc25A contributes to cyclin E1-Cdk2 inhibition at senescence in human mammary epithelial cells. Oncogene 19, 5314–5323.

Sandhu, C., Peehl, D.M., Slingerland, J., 2000b. p16INK4A mediates cyclin dependent kinase 4 and 6 inhibition in senescent prostatic epithelial cells. Cancer Res. 60, 2616–2622.

Serrano, M., Hannon, G.J., Beach, D., 1993. A new regulatory motif in cell cycle control causing specific inhibition of cyclin D/Cdk4. Nature 366, 704–707.

Serrano, M., Lee, H., Chin, L., Cordon-Cardo, C., Beach, D., DePinho, R.A., 1996. Role of the INK4a locus in tumor suppression and cell mortality. Cell 85, 27–37.

Shay, J.W., 1995. Aging and cancer: are telomeres and telomerase the connection? Mol. Med. Today 1, 378–384.

Shay, J.W., Pereira-Smith, O.M., Wright, W.E., 1991a. A role for both RB and p53 in the regulation of human cellular senescence. Exp. Cell Res. 196, 33–39.

Shay, J.W., Wright, W.E., Werbin, H., 1991b. Defining the molecular mechanisms of human cell immortalization. Biochim. Biophys. Acta 1072, 1–7.

Shay, J.W., Wright, W.E., Brasiskyte, D., Van der Haegen, B.A., 1993. E6 of human papillomavirus type 16 can overcome the M1 stage of immortalization in human mammary epithelial cells but not in human fibroblasts. Oncogene 8, 1407–1413.

Sheaff, R.J., Groudine, M., Gordon, M., Roberts, J.M., Clurman, B.E., 1997. Cyclin E-CDK2 is a regulator of p27Kip1. Genes Dev. 11, 1464–1478.

Sherr, C.J., 1994. G1 phase progression: cycling on cue. Cell 79, 551–555.

Sherr, C.J., Roberts, J.M., 1995. Inhibitors of mammalian G1 cyclin-dependent kinases. Genes Dev. 9, 1149–1163.

Slingerland, J.M., Pagano, M., 1997. Regulation of the cell cycle by the ubiquitin pathway. In: M. Pagano (Ed.), Cell Cycle Control in Higher Eukaryotes. Springer Verlag Publishers.

Slingerland, J.M., Hengst, L., Pan, C.-H., Alexander, D., Stampfer, M.R., Reed, S.I., 1994. A novel inhibitor of cyclin-Cdk activity detected in Transforming Growth Factor β-arrested epithelial cells. Mol. Cell. Biol. 14, 3683–3694.

Solomon, M., Lee, T., Kirschner, M., 1992. Role of phosphorylation in p34cdc2 activation: identification of an activating kinase. Mol. Biol. Cell 3, 13–27.

Stein, G.H., Beeson, M., Gordon, L., 1990. Failure to phosphorylate the retinoblastoma gene product in senescent human fibroblasts. Science 249, 666–669.

Stein, G.H., Drullinger, L.F., Soulard, A., Dulic, V., 1999. Differential roles for cyclin-dependent kinase inhibitors p21 and p16 in the mechanisms of senescence and differentiation in human fibroblasts. Mol. Cell. Biol. 19, 2109–2117.

Stevaux, O., Dyson, N.J., 2002. A revised picture of the E2F transcriptional network and RB function. Curr. Opin. Cell Biol. 14(6), 684–691.

Strausfeld, U., Labbe, J.-C., Fesquet, D., Cavadore, J.-C., Picard, A., Sadhu, K., Russell, P., Doree, M., 1991. Dephosphorylation and activation of a p34cdc2/cyclin B complex in vitro by human CDC25 protein. Nature 351, 242–245.

Terada, Y., Tatsuka, M., Jinno, S., Okayama, H., 1995. Requirement for tyrosine phosphorylation of Cdk4 in G1 arrest induced by ultraviolet irradiation. Nature 376, 358–362.

Tsvetkov, L.M., Yeh, K.H., Lee, S.J., Sun, H., Zhang, H., 1999. p27(Kip1) ubiquitination and degradation is regulated by the SCFSkp2 complex through phosphorylated thr187 in p27. Curr. Biol. 9, 661–664.

Vaziri, H., Benchimol, S., 1998. Reconstitution of telomerase activity in normal human cells leads to elongation of telomeres and extended replicative life span. Curr. Biol. 8, 279–282.

Vojta, P.J., Barrett, J.C., 1995. Genetic analysis of cellular senescence. Biochim. Biophys. Acta 1242, 29–41.

Waga, S., Hannon, G.J., Beach, D., Stillman, B., 1994. The p21 inhibitor of cyclin-dependent kinases controls DNA replication by interaction with PCNA. Nature 369, 574–578.

Wang, J., Xie, L.Y., Allan, S., Beach, D., Hannon, G.J., 1998. Myc activates telomerase. Genes Dev. 12, 1769–1774.

Willems, A.R., Lanker, S., Patton, E.E., Craig, K.L., Nason, T.F., Mathias, N., Kobayashi, R., Wittenberg, C., Tyers, M., 1996. Cdc53 targets phosphorylated G1 cyclins for degradation by the ubiquitin proteolytic pathway. Cell 86, 453–463.

Wright, W.E., Shay, J.W., 1992. The two-stage mechanism controlling cellular senescence and immortalization. Exp. Gerontol. 27, 383–389.

Wu, K.J., Grandori, C., Amacker, M., Simon-Vermot, N., Polack, A., Lingner, J., Dalla-Favera, R., 1999. Direct activation of TERT transcription by c-MYC. Nat. Genet. 21, 220–224.

Yang, J., Bardes, E.S., Moore, J.D., Brennan, J., Powers, M.A., Kornbluth, S., 1998. Control of cyclin B1 localization through regulated binding of the nuclear export factor CRM1. Genes Dev. 12, 2131–2143.

Zhu, J., Wang, H., Bishop, J.M., Blackburn, E.H., 1999. Telomerase extends the lifespan of virus-transformed human cells without net telomere lengthening. Proc. Natl. Acad. Sci. USA 96, 3723–3728.

Zindy, F., Quelle, D.E., Roussel, M.F., Sherr, C.J., 1997. Expression of the p16INK4a tumor suppressor versus other INK4 family members during mouse development and aging. Oncogene 15, 203–211.

**Advances in
Cell Aging and
Gerontology**

Protein phosphorylation in T-cell signaling: effect of age

Bulbul Chakravarti* and Deb N. Chakravarti

Keck Graduate Institute of Applied Life Sciences, 535 Watson Drive, Claremont, CA 91711, USA.
Tel.: +1-909-607-9525; fax: +1-909-607-8086,
E-mail address: bulbul_chakravarti@kgi.edu

Contents

Abstract. Aging is a natural biological phenomenon, which is the inevitable fate of all living organisms and is characterized by functional decline of different vital parts of the body including the immune system. Different arms of the immune system are modulated due to aging in humans as well as animal model systems; among these the changes of the T-cell functions have been studied most extensively. In recent years, the biochemical events implicated with the signal transduction cascade responsible for transmission of signal from the cell surface to the nucleus have been studied in detail in a wide variety of cellular systems, including lymphocytes. It is now evident that multiple steps of the signal transduction cascade, including protein phosphorylation and activities of different enzymes, are adversely affected due to aging. This chapter summarizes our current knowledge on two major effector pathways involved in intracellular transmission of T-cell receptor (TCR) signals, one of which is initiated by phospholipase C gamma-1 (PLCγ-1) that leads to the activation of calcium-dependent kinases and phosphatases, and the other one is initiated by Ras-GTP-binding protein that leads to the activation of mitogen-activated protein kinase (MAPK) cascade involving several dual specificity kinases and phosphatases. Age-associated modulation of these effector systems will be discussed in the context of the immune deficiency of T-cells. Such knowledge can provide useful information in designing intervention strategies to improve the immune function of the elderly.

Advances in Cell Aging and Gerontology, vol. 16, 35–56
DOI: 10.1016/S1566-3124(04)16003-3

1. Introduction

T-cells play a central role in the host defense mechanism against infection. However, immune functions of T-cells become impaired due to aging (for references, see Chakravarti and Abraham, 1999; Chakravarti, 2001; Pawelec et al., 2002). Several studies have clearly demonstrated that both early and late events of T-cell activation in humans as well as in animals become impaired with age. In addition, multiple steps of the signal transduction cascade become defective, which may have cumulative effects on the age-related immune deficiency of T-cells. However, the biochemical basis of such alterations of immune function is not fully understood. Due to steadily increasing elderly population in our society, it is important to delineate the basis of the immune deficiency due to aging with the ultimate goal of providing cost effective, improved health care to the elderly. Compared to B-cells and monocytes, more studies have been carried out on the effect of age on protein phosphorylation and related enzymes in T-cells and will be described below.

2. TCR and protein phosphorylation

The T-cell receptor (TCR) is a multisubunit complex composed of at least six different gene products – Tiα and Tiβ chains; the CD3 γ, δ, and ε chains; and a ζ chain (Chan et al., 1994). Polymorphic and disulfide-linked heterodimer α, β subunits are ligand-binding components which belong to the immunoglobulin gene superfamily, have short cytoplasmic domains of five amino acids, and are noncovalently associated with invariant CD3 and ζ chains. The CD3 chain has an extracellular immunoglobulin-like domain and a large cytoplasmic domain consisting of 40–80 amino acids and is noncovalently associated with the ζ chain. The ζ chain is also a transmembrane protein with a nine-amino acid extracellular domain, a single transmembrane domain, and a cytoplasmic domain of 113 amino acids. Although the ζ chain exists primarily as a homodimer, heterodimeric forms have also been reported. Cytoplasmic domains of the subunits of CD3 antigen and ζ chain regulate TCR-mediated signal transduction across the T-cell membrane. The immunoreceptor tyrosine-based activation motif (ITAM) appears tandemly three times in the TCR ζ chain and once in each of the CD3 subunits, and contains a pair of Tyr–X_{aa}–X_{aa}–Leu/Ile phosphoacceptor sites, separated by a linker region of approximately 11 residues (Weiss, 1993). Simultaneous phosphorylation of both tyrosine residues of ITAMs is essential for efficient signal transduction. The enzymes p56lck and p59fyn, which belong to the Src family of tyrosine kinase, are believed to mediate phosphorylation of ITAMs in T-cells. This results in the recruitment and activation of ZAP-70, which is a member of the Syk family of tyrosine kinase (van Oers and Weiss, 1995). As described above, multiple copies of the above motif are present in the TCR, which link the TCR to the effector function. Since only a few hundred receptors are involved following any antigenic stimulation, it is logical to speculate that amplification of the signal following TCR occupancy by its ligand is necessary, and can be facilitated by the multiplicity of ITAM sequences.

The TCR can recognize the antigens processed by the antigen-presenting cells (APCs) in conjunction with the major histocompatibility complex (MHC). It has been shown that the engagement of the TCR by an appropriate MHC–peptide complex is not sufficient for the activation of T-cells since an additional signal provided by accessory molecules present on the surface of the APCs is also necessary. The major T-cell costimulatory pathway involves CD28, the ligands being B7-1 and B7-2 molecules present on APCs (Jenkins et al., 1987). In addition, other T-cell surface molecules, e.g., CD4 and CD8 coreceptors, also play important roles in antigen-driven T-cell activation by binding with monomorphic regions of class II and class I molecules, respectively, thereby helping to stabilize the intercellular complex (MHC–peptide–TCR–CD4/CD8) between the APCs and the antigen-specific T-cells. Protein tyrosine phosphorylation plays an important role in transmembrane signaling following ligation of these receptors. Depending upon the nature of stimulation of TCR, a variety of biological responses can follow which include proliferation, anergy (i.e., unresponsiveness to subsequent stimulation), as well as apoptosis or programmed cell death (for references, see review by Alberola-Ila et al., 1997).

As mentioned earlier, the three different nonreceptor protein tyrosine kinases, which play a crucial role in the signal transduction pathway, are p59fyn, p56lck, and ZAP-70. While p59fyn and p56lck belong to the src family of tyrosine kinase, ZAP-70 belongs to the syk family of tyrosine kinase. The enzyme p56lck, the first among these three kinases to be identified, interacts physically and functionally with the CD4 and CD8 coreceptors, and is expressed exclusively on lymphoid cells and predominantly in T lymphocytes (Abraham and Veillette, 1991). The catalytic activity of p56lck is primarily regulated by phosphorylation of its negative regulatory domain by another src family kinase csk and is dephosphorylated by a protein tyrosine phosphatase CD45 (Mustelin et al., 1989; Nada et al., 1993). The enzyme p59fyn is expressed predominantly in tissues of neuronal and hematopoietic origin, and ZAP-70 is expressed solely in T-cells and natural killer cells (for references, see review by Chan et al., 1994). The function of these src family kinases is to phosphorylate membrane-associated ITAMs following ligation of TCR, which in turn leads to recruitment and binding of ZAP-70 with the ζ chain. Although p59fyn cannot substitute p56lck completely for its function, it plays a crucial role in mature thymocytes. Binding of ZAP-70 with ITAMs may orient the kinase with respect to p56lck. Following phosphorylation by p56lck and/or p59fyn, the activity of ZAP-70 is increased. Interestingly, some studies have pointed out that ZAP-70 can phosphorylate a site within the Src-homology 2 (SH2) domain of p56lck, which can also increase its activity. The interactions between the src family kinase and ZAP-70 have been reviewed by van Oers and Weiss (1995).

Apart from these kinases, a key enzyme in the signal transduction pathway is PLCγ-1, which is recruited to the membrane probably via interaction with an adaptor protein. Following tyrosine phosphorylation, it becomes activated and can hydrolyze membrane-bound phospholipid, phosphatidyl inositol 4,5-biphosphate (PIP2) into inositol 1,4,5-triphosphate (IP3) and diacyl glycerol (DAG).

IP3 production can lead to an increase in Ca^{++}/calmodulin-dependent serine phosphatase (calcineurin). The function of calcineurin is to activate the family of NF-AT transcription factor, which can increase IL-2 gene expression. Apart from calcineurin, there are other calcium/calmodulin-dependent enzymes present in T-cells as well as other cell types (Jensen et al., 1991; Wegner et al., 1992; Hanissian et al., 1993) such as the multifunctional calcium/calmodulin-dependent protein kinase type IV (CaMK-IV), which can phosphorylate the nuclear protein, cAMP-response element-binding protein (CREB), on a serine residue and possibly contributes to increased transcription of immediate early genes containing CRE regulatory sequences (Matthews et al., 1994; Miranti et al., 1995). The other product of PIP2 hydrolysis is DAG, which can activate protein kinase C (PKC). PKC can stimulate accumulation of GTP-bound Ras and hence its downstream effector molecules, which include Raf-1 – a mitogen-activated protein kinase kinase kinase (MAPKKK), MEK – a mitogen-activated protein kinase kinase (MAPKK), and ERK (extracellular response kinase) – a mitogen-activated protein kinase (MAPK). However, it has been reported that TCR-induced Ras activation is not dependent on PKC activation (Izquierdo et al., 1994). The ultimate target molecules of Ca^{++}/calcineurin pathway as well as Ser/Thr kinase cascades (MAPK cascades) are the different nuclear transcription factors necessary for cytokine gene transcription. Although the role of Ras-GTP-binding protein in lymphocyte activation is well accepted, the mechanism by which ligand binding to the TCR/CD3 complex activates Ras is only partly understood. However, inhibition of GTPase-activating proteins (GAPs) as well as activation of guanine-exchange proteins and some adaptor molecules are believed to play important roles (Acuto and Cantrell, 2000; Genot and Cantrell, 2000; Myung et al., 2000). In T lymphocytes, the downstream events of Ras-GTP activation has focused on MAPK cascade, which consists of serine/threonine or dual-specificity kinases. As mentioned above, activation of Ras leads to the sequential activation of Raf-1, MEK, and ERK. In T-cells at least two different isoforms of ERK are present – ERK1 and ERK2, which can be activated in response to stimulation (Whitehurst et al., 1992). When ERKs become activated, they can translocate to the nucleus and regulate the phosphorylation of transcription factors which control the transcriptional activation of immediate early genes, e.g., c-myc, c-fos, and c-jun (Pulverer et al., 1991; Marais et al., 1993). In essence, there are two major effector pathways involved in intracellular transmission of TCR signals similar to those identified in other systems. Initiation of one of those two pathways depends on the activation of PLCγ-1, which regulates calcium-dependent kinases and phosphatases. The second effector system requires the activation of the small GTP-binding molecule Ras and involves several serine/threonine kinases and phosphatases, which can initiate as well as terminate the activation of ERK/MAPK pathway. The ultimate targets of both of these pathways are different transcription factors involved in the cytokine gene transcription.

In addition to ERK, two more members of the MAPK family, which play important roles in T-cell signaling include c-Jun N-terminal kinase (JNK) and p38 MAP kinase (for references, see Rincon et al., 2000). While ERK is normally associated with proliferation and induced by growth factors, the above two MAP kinases are involved with differentiation and cell death, and are induced by cytokines and cellular stress.

Apart from the enzymes mentioned above, a number of polypeptides, which are involved in positive or negative regulation of T-cell signaling pathway, but lack any enzymatic activity, have been identified. They possess interaction domains for several molecules of the T-cell activation pathway. The characteristic structural features of the adaptor molecules include SH2 and phosphotyrosine-binding domain which interact with phosphorylated tyrosine residues in the context of adjacent carboxy- or amino-terminal residues; SH3 domain which interacts with proline-rich sequences; WW domain which is involved with consensus proline–tyrosine or proline–leucine motifs; and plecstrin homology domain which interacts with specific phospholipids (for references, see Myung et al., 2000). These molecules play an important role in propagating signal from ligated TCRs to the nucleus, and some of them are LAT (linker for activated T-cells), Grb2, Shb, Gad, SLP-76 (SH2 domain containing leukocyte protein – 76 kDa), Lnk, p120cbl, Nck, etc. Other protein molecules, e.g., Vav (guanine nucleotide-exchange factor for Rho/Rac molecules) and Sos (guanine nucleotide-exchange factor for Ras molecule) also play important roles in the TCR-signaling pathway. Among the adaptor proteins, LAT, which is present constitutively in the glycolipid-enriched microdomain (GEM), plays a key role. It becomes phosphorylated by ZAP-70 following TCR activation. The phosphorylated LAT helps to recruit various SH2 domain containing proteins to the plasma membrane including PLCγ-1, 85 kDa subunit of phosphoinositide-3-kinase, Grb2 and its family member Gad. Gad constitutively associates with SLP-76, which can also be phosphorylated by ZAP-70. Phosphorylated SLP-76 possibly associates with SH2 domains of Vav and Nck and forms SLP-76–Vav–Nck–Pak1 (p-21-activated protein kinase 1) complex, which may play an important role in cytoskeletal reorganization. Phosphorylated SLP-76 can also bind SLP-76-associated protein (SLAP) and Itk (a Tec family kinase). Itk can subsequently phosphorylate and further activate PLCγ-1. LAT activation also plays important role in ERK activation by forming LAT–Grb2–Sos complex. Recently, it has been suggested that LAT–Grb2–Sos complex is not enough for complete activation of Ras – in fact, PLCγ-1 is also needed (Jordan et al., 2003). Although the exact role of the large group of adaptor molecules identified so far is still under investigation, it possibly facilitates selective transduction of the TCR signal through different effector pathways (for references, see review by Alberota-Ila et al., 1997; Myung et al., 2000; Jordan et al., 2003).

During T-cell activation, cytokines are produced which play important roles in cellular proliferation, differentiation, and apoptosis, a form of programmed cell death. The cytokines bind to their receptors on the cell surface and transduce signal to the nucleus with the help of two different groups of molecules – (i) Janus kinases (JAKs) (JAK-1, -2, -3, and Tyk 2) and (ii) Signal Transducer and Activator of Transcription (STAT) proteins (STAT-1, -2, -3, -4, -5, -6) (for references, see review by Aaronson and Horvath, 2002; Kisseleva et al., 2002). The binding of the cytokines leads to the dimerization of the cytokine receptors followed by the activation of JAKs, a group of tyrosine kinases which associate with the cytokine receptors; activated JAKs can phosphorylate various STAT proteins at tyrosine residues leading to their dimerization and translocation to the nucleus. The activated STAT molecules can bind to the appropriate response elements of the promoter region of specific genes to be transcribed.

Recently, several studies have indicated the presence of membrane rafts in the T-cell membranes, also known as GEMs (glycolipid-enriched microdomains), which are detergent resistant and enriched in sphingolipid and cholesterol, and play a key role in T-cell signal transduction pathway. It has been observed that enrichment of GEMs (which is suggested to concentrate enzymes and substrates of the T-cell signal-transduction pathway and lower the negative regulatory molecules such as tyrosine phosphatase CD45) is necessary at the TCR/APC site for efficient T-cell activation (for references, see Langlet et al., 2000; Tamir et al., 2000; Wang and Eck, 2003). The role of GEMs in T-cell activation is currently under investigation by different groups of investigators which will help to further clarify the possible timing and mechanism of action of different kinases and their substrates, adaptor molecules and other key enzymes of the T-cell signal transduction pathway.

Figure 1 summarizes schematically the T-cell activation pathway. The asterisks denote the enzymes or proteins whose phosphorylation and/or activities or intracellular translocation are altered due to aging.

3. Aging and protein phosphorylation in T-cells

Beginning in 1990, a series of studies have reported the decline of protein phosphorylation ability of the T-cells with aging in rodents as well as humans. However, the majority of such studies have been carried out in mice. Defects have been observed at various levels, such as different adaptor molecules and enzymes including PLCγ-1, tyrosine kinases as well as serine/threonine kinases and phosphatase in humans and/or rodents. However, the molecular identities and physiological functions of some of the proteins with impaired phosphorylation due to aging are not known yet. This chapter will summarize separately the defects observed in rodents and humans.

3.1. In rodents

Initial studies on the age-associated defect in the protein phosphorylation have been carried out in mice. Apart from a few recent reports in rats (Pahlavani et al., 1998; Pahlavani and Vargas, 1999, 2000; Li et al., 2000), all other studies in rodents have used mice. Patel and Miller (1992) have observed age-associated defects in mitogen-induced protein phosphorylation in murine T lymphocytes (Fig. 2).

Lysates prepared from T-cells following stimulation with mitogens (anti-CD3, concanavalin A [ConA], and phorbol myristate acetate [PMA] plus ionomycin) as well as nonmitogenic stimulation (PMA or ionomycin used separately) have shown a gradual downregulation of phosphorylation of proteins of a wide range of

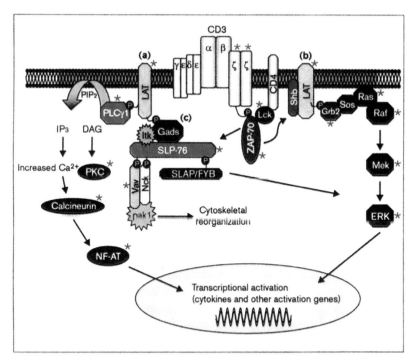

Fig. 1. T-cell activation pathway and its impairment due to aging. Following TCR engagement (here on a CD4 cell as an example), Lck is activated leading to CD3ζ and ZAP-70 phosphorylation. Activated ZAP-70 phosphorylates SLP-76 and LAT, initiating Ras/MAPK and PLCγ-1 signaling cascades. Tyrosine phosphorylation of LAT, which is constitutively located in GEMs (illustrated by green membrane section), (a) recruits PLCγ-1 to the membrane, placing PLCγ-1 in close proximity with its substrate, PIP2. PIP2 is hydrolyzed to IP3 and DAG, which leads to increases of cytosolic-free calcium and activation of PKC. (b) Tyrosine-phosphorylated LAT also binds to the Grb2 SH2 domain, which recruits Sos to the plasma membrane thus initiating Ras/ERK activation. Since Gads constitutively associates with SLP-76, (c) the SLP-76–Gads complex may be recruited to LAT leading to Ras/MAPK and PLCγ-1 activation. Tyrosine phosphorylated SLP-76 also associates with Vav, thus promoting the formation of a SLP-76–Vav–Nck–Pak 1 complex, which may be important for the regulation of cytoskeletal rearrangements. SLP-76 additionally binds to the SH2 domain of Itk, which subsequently phosphorylates and further activates PLCγ-1. Other signaling mediators not depicted include activation of Rac 1 and other MAPKs, including JNK. The asterisks denote the different components of the T-cell signal transduction pathway whose activities and/or tyrosine phosphorylation or translocation are impaired due to aging. (Reprinted with minor modification from Current Opinion in Immunology, Volume 12, Peggy S. Myung, Nancy J. Boerthe and Gary A. Koretzky, "Adapter proteins in lymphocyte antigen-receptor signaling", pp. 256–266, 2000, with permission from Elsevier.)

molecular weights with age. The effect has been observed in virgin as well as memory T-cells. However, the molecular identity or physiological significance of these substrates has not been characterized. Although analysis of phosphoamino acid was not carried out, the higher proportion of serine–threonine-specific phosphoproteins compared to tyrosine-specific phosphoproteins and responsiveness of most of the phosphoproteins to PMA and calcium ionophore, which activate

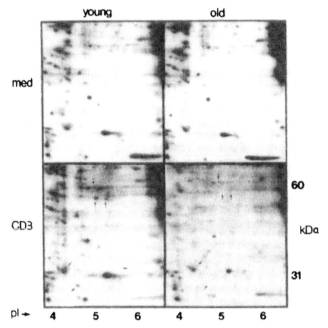

Fig. 2. Age-related declines in protein phosphorylation after stimulation with anti-CD3. Splenic T-cells were labeled with ^{32}P and stimulated for 10 min with anti-CD3; lysates were separated by 2-D electrophoresis and then analyzed by autoradiography. Arrows indicate some of the phosphoproteins whose phosphorylation level increases in response to anti-CD3 in young cells, but not in old cells. (Reprinted from European Journal of Immunology, Volume 22, Hiren R. Patel and Richard A. Miller, "Age-associated changes in mitogen-induced protein phosphorylation in murine T lymphocytes", pp. 253–260, 1992, with permission from Wiley-VCH.)

serine and threonine kinases preferentially were observed. This suggests that aging leads to an alteration of serine- and threonine-specific kinases and phosphatases. Recent studies have confirmed this observation and will be discussed later in this section.

Although phosphorylation of tyrosine residues accounts for only a small percentage of the total phosphate associated with proteins *in vivo* (Hunter and Cooper, 1985), it plays a central role in cell proliferation and differentiation and hence in senescence. As a result, a series of studies have been carried out to identify the effect of age on impairment of protein tyrosine phosphorylation in T-cells. Anti-phosphotyrosine monoclonal antibody has been used in those studies to identify tyrosine-phosphorylated proteins. A defect in tyrosine phosphorylation of different proteins following stimulation of unfractionated T-cells with anti-CD3 (120, 80, or 40 kDa) or anti-$\alpha\beta$ TCR (120 and 80 kDa) has been observed (Shi and Miller, 1992, 1993). Interestingly, the effect of age on tyrosine phosphorylation in mouse has been found to vary among T-cell subsets. CD8 cells have shown age-related impairment of all three substrates in response to anti-CD3 and ConA, but only of p80 (and not of p120) following anti-TCR stimulation. The p80 response of CD4 cells has been found to decline with age with all three stimulants while the responses of p120 and p40 have been relatively insensitive to age. The age-dependent decline in CD4 cell phosphorylation of p80 can be correlated with the age-associated shift from naive to memory cells. In fact, memory T-cells from young and old donors generate less-phosphorylated p80 than unfractionated T-cells in response to all three stimulants. Similarly, phosphorylation of p40 in response to anti-CD3 and ConA, and phosphorylation of p120 in response to anti-CD3 in isolated CD4 memory cells from young as well as old mice has been estimated to be lower compared to those observed in unfractionated T-cells. From the above results, it is clear that protein tyrosine kinase-mediated signal transduction pathway differs between CD4 and CD8 subsets as well as between naive and memory cells of CD4 subsets. This may be the underlying cause of difference in activation requirement for the assay of proliferation and cytokine production by different T-cell subsets (Czitrom et al., 1983; Luqman and Bottomly, 1992).

Subsequently, different investigators have reported defects in tyrosine phosphorylation of specific proteins and enzymes of the signal transduction pathway in mice. For example, tyrosine phosphorylation of (i) Shc (Ghosh and Miller, 1995) in anti-CD3-stimulated T-cells, (ii) ζ chain in resting as well as anti-CD3-stimulated T-cells (Garcia and Miller, 1997, 1998), (iii) PLCγ-1 and a 35/36-kDa phosphoprotein which specifically binds to SH2 domain of PLCγ-1 and a possible regulatory protein for activation of PLCγ-1 by its kinases following ligation of CD3 or CD3 × CD4 receptors (Grossmann et al., 1995; Utsuyama et al., 1997) are impaired in aged mice. Phosphorylation of LAT has also been reported to be downregulated due to aging (Tamir et al., 2000). In addition, the amount of PLCγ-1 is lower in T-cells of old mice (Utsuyama et al., 1997). PIP2, which is known to be hydrolyzed into IP3 and DAG, has been found to be present in comparable amount in the cell membranes of young and old mice. However, there are significant decreases in the production of IP3 and DAG due to aging. Since tyrosine kinases such as p59fyn and ZAP-70 are believed to be involved in the activation of PLCγ-1, the effect of age on the tyrosine phosphorylation level of these kinases have also been studied. Age has been found to be associated with a decrease of tyrosine phosphorylation of both p59fyn and ZAP-70. Utsuyama et al. (1997) have observed the impairment in the tyrosine phosphorylation of two different substrates with molecular weights 110 and 75 kDa, which may be identical with those proteins whose molecular weights have been estimated to be 120 and 80 kDa by Shi and Miller (1992, 1993) and are impaired in tyrosine phosphorylation due to aging.

Since activation-induced ζ phosphorylation leads to increase in recruitment and binding of ZAP-70 and other SH2-containing key protein molecules of the signal transduction pathway, the age-associated decline in ζ phosphorylation is expected to cause a decline in its association with ZAP-70 and consequently its kinase activity. On the contrary, a two-fold increase in ZAP-70 association with CD3ζ in resting CD4 T-cells has been observed (Garcia and Miller, 1998). Also, a decline in tyrosine phosphorylation of Shc can impair the activation of Ras pathway since it is implicated with the transmission of an activation signal from the stimulated T-cell receptors to Ras. In fact, Shc has been shown to be phosphorylated and bind to TCR ζ as well as Grb2/Sos complex following TCR stimulation (Ravichandran et al., 1993). Defects in activation in Ras signaling due to aging have been observed by different investigators in rodents and will be discussed below.

Although the majority of the studies indicate impairment of phosphorylation of different substrates due to aging, occasionally higher phosphorylation of certain substrates have also been observed. Utsuyama et al. (1997) have demonstrated higher phosphorylation of a 65-kDa protein in T-cells from old compared to young mice. Earlier, Ghosh and Miller (1995) reported an age-associated increase in tyrosine phosphorylation of Grb2, Shc, and a 30-kDa Grb2-like protein following stimulation via CD4 but not via CD3. This indicates an age-dependent alteration in the manner in which CD4/p56lck complexes are regulated and coupled to other components in the T-cell activation pathway. Further studies will clarify how the age-associated increase in the phosphorylation of these substrates can influence the host immune function.

While the defect in PLCγ-1 can be reflected in the signaling cascade involving calcium-dependent kinases and phosphatases, defects have also been observed in the other arm of the signal transduction pathway, i.e., different serine/threonine kinases of the MAPK pathway following TCR activation. Utsuyama et al. (1997) have reported impairment of tyrosine phosphorylation of p42 MAPK. Subsequently, Kirk and Miller (1998) have demonstrated that aging can lead to two- to four-fold decline in Raf-1 activity, an upstream event of MAPK activation when T-cells from mice are activated via antibodies to CD3 ε chain and CD4, but the kinetics of the enzyme activity does not change due to aging. Studies with mice belonging to the same age group have demonstrated that Raf-1 activation due to CD3/CD4 stimulation is lower in the memory cells compared to the naive cells. However, aging leads to decline in Raf-1 activity in naive subset of CD4 T-cells. In addition, it has been observed that antibodies to the costimulatory molecule CD28 can also activate Raf-1 and an additive effect has been observed when cells are activated via CD28 and CD3. However, even following stimulation via CD28 and CD3, Raf-1 activity from old mice does not reach the level of that from the young mice. Also, a decline in phosphorylation of c-Jun in CD4 T-cells stimulated by CD3/CD4/CD28 cross-linking due to diminished activation of JNK has been reported (Kirk and Miller, 1999; Kirk et al., 1999).

In similar studies with rats, Pahlavani et al. (1998) observed that induction of MAPK and Ras in T-cells stimulated with ConA is significantly lower in those isolated from the old rats compared to those from the young rats. However, contrary to the report in mice (Kirk and Miller, 1999), Pahlavani et al. (1998) have failed to observe any significant alteration in the induction of JNK activity in rats due to aging. Interestingly, the time course of induction of MAPK, JNK, and Ras activities are similar in rats from both age groups. Although the proportion of p44 MAPK has been found to decline with age, there is no age-associated difference in the expression of p42 and p44 MAPKs. They also observed a decline in the activities of p56lck and ZAP-70 due to aging but not of p59fyn. The apparent discrepancy between the effect of age on JNK activation observed by different groups of investigators may be due to the differences in species and/or stimulating agents used. Later, Li et al. (2000) have bypassed receptor proximal events and stimulated rat splenic lymphocytes with phorbol ester (PMA) and calcium ionophore (A23187) as comitogens, and observed impairment in the activation of both ERK and JNK MAPK signal transduction pathways. Since activation of MEK is also inhibited, age-related defect in ERK activity is possibly upstream event of ERK itself. On the other hand, the observed defect in MAPK activation results in decreased activation of downstream events such as c-Jun phosphorylation. Pahlavani and Vargas (1999, 2000) have also observed defects in the other arm of the T-cell signal transduction pathway, i.e., in calcium signaling in rats. When rat splenic cells are stimulated with anti-CD3 antibody or ConA, there is age-associated decline in the activity of calcium/calmodulin-dependent phosphatase calcineurin (dephosphorylates and translocates cytoplasmic NF-AT to the nucleus) and calcium/ calmodulin-dependent kinase CaMK-IV (plays a critical role in calcium signaling). The age-related decline in the activity of calcineurin and CaMK-IV is not due to change in the expression level of these enzymes.

Assembly of signal transduction complex at the site of T-cell/APC interaction is altered due to aging (Yang and Miller, 1999; Eisenbraun et al., 2000; Tamir et al., 2000; Garcia and Miller, 2001). In fact, single-cell analysis using confocal microscopy suggests that naive CD4 T-cells from old mice undergo two different defects in T-cell/APC interaction following peptide-specific activation. Firstly, it leads to approximately two-fold decrease in the proportion of T-cell/APC conjugates that could relocalize different proteins (LAT, ZAP-70, c-Cbl, Vav, Grb-2, PLCγ-1, PKC theta, p59fyn, and p56lck) of the T-cell signal transduction pathway. Secondly, there is a two-fold decrease in the frequency of cytoplasmic NF-AT migration among cells that could generate immune synapses containing LAT, c-Cbl, and PLCγ. In addition, there is a defect in cytoskeletal reorganization even earlier to the localization of protein kinases and their substrates to the TCR/APC sites (Garcia and Miller, 2002).

Recently, it has been observed that defective T-cell priming associated with aging can be reversed by signaling through 4-1BB (CD 137), a member of the TNFR family which is able to provide expansion and survival signals to T-cells (Bansal-Pakala and Croft, 2002).

3.2. In humans

Studies similar to mice have also been carried out in humans. Quadri et al. (1996) have observed age-related defect in tyrosine-specific protein phosphorylation in T-cells activated through various receptors including CD3, CD4, CD8, and IL-2 receptor (IL-2R) suggesting that age-associated defect in tyrosine phosphorylation is not restricted to TCR-mediated signal transduction but affects CD4, CD8, and IL-2R initiated signals as well. Since purified IL-2R bearing cells from elderly donors cannot initiate significant DNA synthesis irrespective of the amount of supplemented

IL-2, it is logical to presume that there is impairment in IL-2R signaling due to aging. Quadri et al. (1996) have observed a defect in tyrosine phosphorylation of a limited number of proteins following IL-2R activation in the lysates of T-cells isolated from elderly donors. This may be the underlying cause of the failure of a high proportion of cells from aged subjects to enter the S phase of the cell cycle (Negoro et al., 1986; Kubbies, 1992).

Chakravarti et al. (1998) also demonstrated an age-associated decline in the tyrosine phosphorylation abilities in human T-cells following stimulation with ConA, phytohemagglutinin [PHA], and anti-CD3 antibody. Contrary to the observation of Shi and Miller (1993) in mice, Chakravarti et al. (1998) failed to observe any difference in the tyrosine phosphorylation ability of CD4 and CD8 T-cells isolated from young and elderly donors. It has been documented that the tyrosine phosphorylation of ZAP-70 is downregulated during aging in unfractionated T-cells [Fig. 3(1)] as well as CD4 and CD8 subsets following nonspecific stimulation with ConA and PHA as well as via CD3 (Chakravarti et al., 1998; Chakravarti and Abraham, 2002).

Although the basis or the effect of the observed decline in tyrosine phosphorylation was not studied by Chakravarti et al. (1998), studies by other investigators (Whisler et al., 1997, 1998, 1999; Guidi et al., 1998; Tinkle et al., 1998; Fulop et al., 1999) have demonstrated a decline in activity and/or tyrosine phosphorylation of p56lck, p59fyn, and ZAP-70 due to aging. Interestingly, the age-associated downregulation of tyrosine phosphorylation of ZAP-70 is not an intrinsic defect and is reversible. When T-cells were incubated for 2 days in serum-free medium and stimulated with anti-CD3 antibody, no age-associated impairment of tyrosine phosphorylation of ZAP-70 was observed (Chakravarti and Abraham, 2002). In fact, when cultured *in vitro* for 2 days, cells from majority of the elderly donors showed a significant increase in *de novo* tyrosine phosphorylation of ZAP-70. However, under the same condition, the young donor cells showed a mixed response, i.e., increase, decrease, or no change in tyrosine phosphorylation of ZAP-70. There was no significant change in the constitutive level of tyrosine phosphorylation for cells from both age groups [Fig. 3(2, A–D)].

A decline in tyrosine phosphorylation of ζ chain following activation with anti-CD3 antibody has also been observed (Whisler et al., 1998, 1999). Since there is a significant age-related defect in the ability of p56lck to associate with CD4 and CD45 (Tinkle et al., 1998), this may be the cause of lower phosphorylation and catalytic activity of p56lck and consequently reduced tyrosine phosphorylation of ZAP-70. Decreased association of p56lck with CD4 may be due to their altered conformation due to aging. In fact, Sharma et al. (1980) have reported altered conformation of proteins due to aging with particular reference to muscle phosphoglycerate kinase in rat. However, no age-associated modulation of the expression of p50csk, another src family kinase that is known to regulate the activity of p56lck has been observed; but whether its activity changes due to aging has not been studied yet (Tinkle et al., 1998). Different investigators have failed to observe an age-related decline in the expression of p59fyn, p56lck, or ZAP-70, which could be correlated with the age-related decline in the activities of these

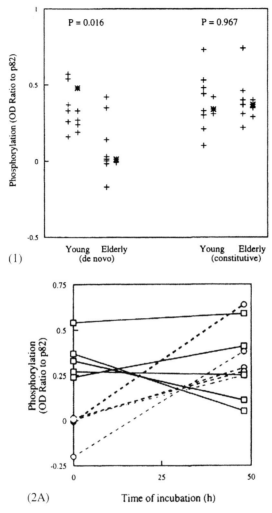

Fig. 3. (1) The effect of age on tyrosine phosphorylation of ZAP-70 in freshly isolated T-cells. Anti-CD3 or IgG1 (control) monoclonal antibody (mAb) was used as the stimulating agent. Following stimulation, cells were lysed and the lysates were separated by SDS-polyacrylamide gel electrophoresis (PAGE) and electrophoretically transferred to nitrocellulose membrane. Anti-ZAP-70 and anti-phosphotyrosine mAbs were used to identify tyrosine-phosphorylated ZAP-70. Tyrosine phosphorylation of ZAP-70 was estimated by staining the membrane with anti-phosphotyrosine mAb and chemiluminescence detection system using autoradiography film. The film was scanned in a laser densitometer and the optical density (OD) was measured. Net tyrosine phosphorylation of ZAP-70 was estimated as the difference of its OD (normalized to p82, a 82-kDa tyrosine-phosphorylated protein which was used as an internal standard; for details, see Chakravarti et al., 1998) in the postnuclear supernatant made from lysed T-cells stimulated in the presence of anti-CD3 mAb and nonspecific murine mAb (IgG1). Asterisks represent the geometric mean values. To obtain the geometric mean value of tyrosine phosphorylation of ZAP-70 for any donor who showed dephosphorylation, the negative OD value was substituted by a low OD value (0.001). (Reprinted from Mechanism of Ageing and Development, Volume 123, Bulbul Chakravarti and George N. Abraham, "Effect of age and oxidative stress on tyrosine phosphorylation of ZAP-70", pp. 297–311, 2002, with permission from Elsevier.) Fig. 3(2) Continued

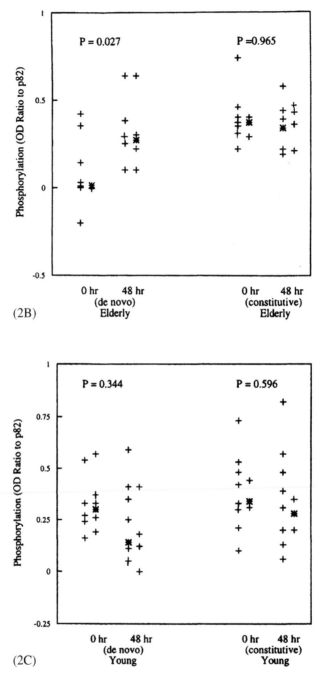

(2B)

(2C)

Fig. 3 (2) Continued

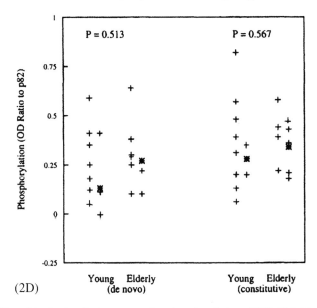

Fig. 3. (2) (A) Effect of *in vitro* incubation on tyrosine phosphorylation of ZAP-70 in T-cells isolated from the elderly (denoted by circles joined by broken lines) and young donors (denoted by squares joined by solid lines). (B) Tyrosine phosphorylation of ZAP-70 in freshly isolated and *in vitro* incubated T-cells purified from the elderly donors. (C) Tyrosine phosphorylation of ZAP-70 in freshly isolated and *in vitro* incubated T-cells purified from young donors. (D) Effect of age on tyrosine-specific phosphorylation of ZAP-70 in *in vitro* incubated T-cells isolated from young and elderly donors. In all of the above figures (B–D), asterisks represent the geometric mean values. To obtain the geometric mean value of tyrosine phosphorylation of ZAP-70, any donor which showed dephosphorylation, the negative OD value was substituted by a low OD value (0.001). T-cells freshly isolated or preincubated for 48 h (for details, see Chakravarti and Abraham, 2002) were stimulated with anti-CD3 or IgG1 (control) mAb. Following stimulation, cells were lysed and the lysates were separated by SDS-PAGE and electrophoretically transferred to nitrocellulose membrane. Anti-ZAP-70 and anti-phosphotyrosine mAbs were used to identify tyrosine-phosphorylated ZAP-70. Tyrosine phosphorylation of ZAP-70 was estimated by staining the membrane with anti-phosphotyrosine mAb and chemiluminescence detection system using autoradiography film. The film was scanned in a laser densitometer and the OD was measured. Net phosphorylation of ZAP-70 was estimated as the difference of its OD (normalized to p82, a 82-kDa tyrosine-phosphorylated protein which was used as an internal standard; for details, see Chakravarti et al., 1998) in the postnuclear supernatant made from lysed T-cells stimulated in presence of anti-CD3 mAb or nonspecific murine mAb (IgG1). (Reprinted from Mechanism of Ageing and Development, Volume 123, Bulbul Chakravarti and George N. Abraham, "Effect of age and oxidative stress on tyrosine phosphorylation of ZAP-70", pp. 297–311, 2002, with permission from Elsevier.)

enzymes. Contrary to the report in mice where decreased tyrosine phosphorylation of ζ chain does not cause impairment of activity of ZAP-70 (Garcia and Miller, 1998), the decline in activity of ZAP-70 in humans due to aging has been correlated with the decreased tyrosine phosphorylation of ζ chain (Whisler et al., 1999). Moreover, the impairment of tyrosine phosphorylation is present in naive as well as memory T-cells. Since no difference has been observed between the activities of phosphatases in cells from the two different age groups, a defect in the regulation of TCR/CD3-coupled protein tyrosine kinases is possibly the basis of age-related defects in the signaling cascade and immune responsiveness in the human T-cells. Another important enzyme of the early phase of T-cell signal transduction pathway is phospholipase C; variation in the expression level of different isoforms of phospholipase C in T-cells has been observed during aging (Di Pietro et al., 2000).

Prior to the studies reported in mice and rats and also more recently, several investigators (Whisler et al., 1996; Liu et al., 1997; Douziech et al., 2002) have observed age-associated decline in the activation of ERK1 and ERK2 as well as MEK in human T-cells stimulated via TCR/CD3 complex. Kinetic analysis has revealed defect in the duration as well as the level of MAPK activation. The down-regulatory effect of age on MEK/MAPK in T-cells of elderly subjects is neither due to reduced expression of these enzymes nor due to failure to activate protein kinase C. Possibly, the defects are not intrinsic to MAPK; rather it is the reflection of the defect in the upstream events of MAPK cascade. Reduction in the activation of JNK and p38 has also been demonstrated in human T-cells due to aging (Liu et al., 1997; Douziech et al., 2002). The decline in the activation of JNK is not due to its altered expression.

Recently, Fulop et al. (2001) have reported age-related alterations in the JAK–STAT signaling pathway that result in decreased tyrosine phosphorylation of STAT 5. They have also suggested a role of cholesterol content of the lymphocyte plasma membrane of the elderly in the regulation of TCR signaling. Aging also leads to a decrease in SLP-76 activation by tyrosine phosphorylation in T-cells after TCR stimulation in humans and may be due to (1) decreased activation of ZAP-70 and/or (2) alteration of the microenvironment of the lipid raft. In fact, the activation of SLP-76 in cells of the elderly was improved following extraction of cholesterol by methyl-β-cyclodextrin (MBCD) (Fulop et al., 2002). Similarly, MBCD induced an increase in the phosphorylation of p56lck, especially in the T-cells of elderly humans (Fulop et al., 2001).

4. Conclusion and future direction

Changes in protein phosphorylation and protein kinase activities are characteristics of normal aging as well as age-related diseases, e.g., osteoporosis, Alzheimer's disease, etc. (for references, see Armbrecht et al., 1993). The present chapter has focused on changes in protein phosphorylation in T-cell signaling due to aging, which may be a leading cause of the observed age-associated impairment of the immune function. Protein kinases can be divided into two major classes – those that phosphorylate tyrosine amino acid residues and those which phosphorylate serine and threonine residues. Although the former group of kinases constitutes only 1% of the total cellular protein kinase activities, they play an indispensable role in cellular growth, differentiation, division, and hence in senescence.

As described above, aging has deleterious effects on the signal transduction pathway in T-cells. Using various stimulating agents, it has been observed that there is downregulation of protein tyrosine phosphorylation of various substrates in humans as well as in rodents. The molecular identities and physiological functions of some of the substrates are known while those for the others need to be identified. Various kinase activities have also been found to be downregulated due to aging, which is possibly implicated with the observed downregulation of protein tyrosine phosphorylation due to aging. While the molecular identities of all these molecules are not yet clear, some of them have been characterized in humans and/or in mouse, such as PLCγ-1, Shc, LAT, SLP-76, ζ chain, ZAP-70, etc. Interestingly, all of them are known to play an important role in the signal transduction pathway. In spite of some discrepancies observed between mice and humans, there is convincing evidence for defects in two major signaling cascades, i.e., one involving PLCγ-1 and the other involving Ras-GTP-binding protein in rodents as well as humans. Also, activation of different kinases such as p56lck, p59fyn, and/or ZAP-70 has been reported to decline with age in humans as well as rodents. A decline in the activity of calcineurin, a calcium/calmodulin-dependent phosphatase, has been observed in rodents due to aging.

It is not yet clear whether the defects observed at multiple steps of distal events of T-cell activation, e.g., proliferation, pattern of cytokine production, etc., are the consequences of several independent age-sensitive changes of the signal transduction pathway or can be correlated with a single proximal event of T-cell activation which is an upstream event of a wide range of responses observed in the signal transduction pathway. However, the observed decline in protein tyrosine phosphorylation may be due to any one or all of the following reasons: (1) since phosphorylation level of any compound depends on the balance between the activities of kinases and phosphatases, change in the protein kinase and protein phosphatase activities due to aging may alter the phosphorylation level of their substrates. Recent studies have provided evidences that there are changes in protein kinase activities during aging. Interestingly, age-associated decline in tyrosine phosphorylation level of ZAP-70 has been found to be reversible. Such studies on the effect of age on phosphatase activities are limited. (2) Lower amount of protein may be present which may be the cause of decline in phosphorylation due to aging. Except for various forms of phospholipase C including that of PLCγ-1, majority of the investigators have reported no change in the expression level of different enzymes of the signal transduction cascade, e.g., p56lck, p59fyn, ZAP-70, JNK, or ERK. (3) A conformational change of the proteins may lead to defective phosphorylation. Although no such data are available yet for the kinases or their substrates, conformational alterations have been observed for muscle phosphoglycerate kinase in old rats (Sharma et al., 1980). Decline in kinase activity during aging may provoke intervention studies using compounds, which are known to modulate protein kinase or phosphatase activities *in vitro*. However, instead of using any such nonspecific modulator, it will be helpful to design specific drugs, which will act on selected kinases or phosphatases of interest.

While it has been shown unambiguously by different investigators that there are abnormalities in protein phosphorylation during aging, the biochemical basis of the defect is not yet clear. However, one possibility may be oxidative stress. Although oxidative stress has been implicated with various aspects of biological aging, its exact role in immunological aspects of aging is not yet clear and certainly requires thorough investigation. Initial studies in mice and rats have shown that the immune functions of young animals are damaged more severely than the old subjects under oxidative stress due to the presence of higher preponderance of naive cells, which are more sensitive to oxidative stress-induced signal compared to memory cells (Lahdenpohja and Hurme, 1996; Lohmiller et al., 1996; Pahlavani and Harris, 1998). In contrast, in a study on the effect of age and oxidative stress on cellular proliferation in humans, Franceschi et al. (1990) observed higher vulnerability of peripheral blood mononuclear cells isolated from elderly donors toward oxidative stress compared to those from young donors. However, we observed no significant difference on the effect of oxidative stress on protein tyrosine phosphorylation ability of T-cells isolated from young or elderly donors (Chakravarti and Abraham, 2002). Certainly this area requires thorough investigation in the future. If the age-related downregulation of protein phosphorylation can be correlated with oxidative stress, intervention strategies using different antioxidants can be used to see if that improves the protein phosphorylation ability and eventually the immune functions of the elderly population.

Another future direction of research on age-associated modification of the T-cell signal transduction pathway should utilize the rapidly evolving proteomics technology. This will help identification of new molecules of the signal transduction pathway, which may or may not be altered due to aging. In addition, detailed structural analysis of any molecule of interest will provide useful information for future drug design to overcome the possible adverse effect of any structural modification.

Acknowledgments

We would like to thank the Keck Graduate Institute for providing support for preparation of the manuscript.

References

Aaronson, D.S., Horvath, C.M., 2002. A road map for those who know JAK-STAT. Science 296, 1653–1655.

Abraham, N., Veillette, A., 1991. The lymphocyte-specific tyrosine protein kinase p56lck. Cancer Invest. 9, 455–463.

Acuto, O., Cantrell, D., 2000. T cell activation and the cytoskeleton. Annu. Rev. Immunol. 18, 165–184.

Alberola-Ila, J., Takaki, S., Kerner, J.D., Perlmutter, R.M., 1997. Differential signaling by lymphocyte antigen receptors. Annu. Rev. Immunol. 15, 125–154.

Armbrecht, H.J., Nemani, R.K., Wongsurawat, N., 1993. Protein phosphorylation: changes with age and age-related diseases. J. Am. Geriatr. Soc. 41, 873–879.

Bansal-Pakala, P., Croft, M., 2002. Defective T cell priming associated with aging can be rescued by signaling through 4-1BB (CD137). J. Immunol. 169, 5005–5009.

Chakravarti, B., 2001. T-cell signaling – effect of age. Exp. Gerontol. 37, 33–39.

Chakravarti, B., Abraham, G.N., 1999. Aging and T-cell mediated immunity. Mech. Ageing Dev. 108, 183–206.

Chakravarti, B., Abraham, G.N., 2002. Effect of age and oxidative stress on tyrosine phosphorylation of ZAP-70. Mech. Ageing Dev. 123, 297–311.

Chakravarti, B., Chakravarti, D.N., Devecis, J., Seshi, B., Abraham, G.N., 1998. Effect of age on mitogen induced protein tyrosine phosphorylation in human T cell and its subsets: down-regulation of tyrosine phosphorylation of ZAP-70. Mech. Ageing Dev. 104, 41–58.

Chan, A.C., Desai, D.M., Weiss, A., 1994. The role of protein tyrosine kinases and protein tyrosine phosphatases in T cell antigen receptor signal transduction. Annu. Rev. Immunol. 12, 555–592.

Czitrom, A.A., Sunshine, G., Reme, T., Ceredig, R., Glasebrook, A.L., Kelso, A., MacDonald, H.R., 1983. Stimulator cell requirements for allospecific T cell subsets: specialized accessory cells are required to activate helper but not cytotoxic T lymphocyte precursors. J. Immunol. 130, 546–550.

Di Pietro, R., Miscia, S., Cataldi, A., Rana, R., 2000. Age-dependent variations in the expression of PLC isoforms upon mitogenic stimulation of peripheral blood T cells from healthy donors. Br. J. Haematol. 111, 1209–1214.

Douziech, N., Seres, I., Larbi, A., Szikszay, E., Roy, P.M., Arcand, M., Dupuis, G., Fulop, T., 2002. Modulation of human lymphocyte proliferative response with aging. Exp Gerontol. 37, 369–387.

Eisenbraun, M.D., Tamir, A., Miller, R.A., 2000. Altered composition of the immunological synapse in an anergic, age-dependent memory T cell subset. J. Immunol. 164, 6105–6122.

Franceschi, C., Cossarizza, A., Troiano, L., Salati, P., Monti, D, 1990. Immunological parameters in aging: studies on natural immunomodulatory and immunoprotective substances. Int. J. Clin. Pharm. Res. 10, 53–57.

Fulop, T., Gagne, D., Gouler, A.-C., Desgeorges, S., Lacombe, G., Arcand, M., Dupuis, G., 1999. Age-related impairment of p56lck and ZAP-70 activities in human T lymphocytes activated through the TCR/CD3 complex. Exp. Gerontol. 34, 197–216.

Fulop, T., Douziech, N., Goulet, A.C., Desgeorges, S., Linteau, A., Lacombe, G., Dupuis, G., 2001. Cyclodextrin modulation of T lymphocyte signal transduction with aging. Mech. Ageing Dev. 122, 1413–1430.

Fulop, T., Douziech, M., Larbi, A., Dupuis, G., 2002. The role of lipid rafts in T lymphocyte signal transduction with aging. Ann. N.Y. Acad. Sci. 973, 302–304.

Garcia, G.G., Miller, R.A., 1997. Differential tyrosine phosphorylation of zeta chain dimers in mouse CD4 T lymphocytes: effect of age. Cell Immunol. 175, 51–57.

Garcia, G.G., Miller, R.A., 1998. Increased ZAP-70 association with CD3 zeta in CD4 T cells from old mice. Cell Immunol. 190, 91–100.

Garcia, G.G., Miller, R.A., 2001. Single cell analyses reveal two defects in peptide-specific activation of naive T cells from aged mice. J. Immunol. 166, 3151–3157.

Garcia, G.G., Miller, R.A., 2002. Age-dependent defects in TCR-triggered cytoskeletal rearrangement in CD4+ T cells. J. Immunol. 169, 5021–5027.

Genot, E., Cantrell, D.A., 2000. Ras regulation and function in lymphocytes. Curr. Opin. Immunol. 12, 289–294.

Ghosh, J., Miller, R.A., 1995. Rapid tyrosine phosphorylation of Grb2 and Shc in T cells exposed to anti-CD3, anti-CD4, and anti-CD45 stimuli: differential effects of aging. Mech. Ageing Dev. 80, 171–187.

Grossmann, A., Rabinovitch, P.S., Kavanagh, T.J., Jinneman, J.C., Gilliland, L.K., Ledbetter, J.A., Kanner, S.B., 1995. Activation of murine T cells via phospholipase-C-gamma 1-associated protein tyrosine phosphorylation is reduced with aging. J. Gerontol. A. Biol. Sci. Med. Sci. 50, B205–B212.

Guidi, L., Antico, L., Bartoloni, C., Costanzo, M., Errani, A., Tricerri, A., Vangeli, M., Doria, G., Gatta, L., Goso, C., Mancino, L., Frasca, D., 1998. Changes in the amount and level of phosphorylation of p56 (lck) in PBL from aging humans. Mech. Ageing Dev. 102, 177–186.

Hanissian, S.H., Frangakis, M., Bland, M.M., Jawahar, S., Chatila, T.A., 1993. Expression of a Ca2+/calmodulin-dependent protein kinase, CaM kinase-Gr, in human T lymphocytes: regulation of kinase activity by T-cell receptor signaling. J. Biol. Chem. 268, 20055–20063.

Hunter, T., Cooper, J.A., 1985. Protein tyrosine kinases. Annu. Rev. Biochem. 54, 897–930.

Izquierdo, M., Leevers, S.J., Williams, D.H., Marshall, C.J., Weiss, A., Cantrell, D., 1994. The role of protein kinase C in the regulation of extracellular signal-regulated kinase by the T cell antigen receptor. Eur. J. Immunol. 24, 2464–2468.

Jenkins, M.K., Pardoll, D.M., Mizuguchi, J., Quill, H., Schwartz, R.H., 1987. T-cell unresponsiveness in vivo and in vitro: fine specificity of induction and molecular characterization of the unresponsive state. Immunol. Rev. 95, 113–135.

Jensen, K.F., Ohmstede, C.A., Fisher, R.S., Sahyoun, N., 1991. Nuclear and axonal localization of $Ca2+$/calmodulin dependent protein kinase type Gr in rat cerebellar cortex. Proc. Natl. Acad. Sci. 88, 2850–2853.

Jordan, M.S., Singer, A.L., Koretzky, G.A., 2003. Adaptors as central mediators of signal transduction in immune cells. Nature Immunol. 4, 110–116.

Kirk, C.J., Freilich, A.M., Miller, R.A., 1999. Age-related decline in activation of JNK by TCR- and CD28-mediated signals in murine T-lymphocytes. Cell Immunol. 197, 75–82.

Kirk, C.J., Miller, R.A., 1998. Analysis of Raf-1 activation in response to TCR activation and costimulation in murine T-lymphocytes: effect of age. Cell Immunol. 190, 33–42.

Kirk, C.J., Miller, R.A., 1999. Age-sensitive and insensitive pathways leading to JNK activation mouse CD4(+) T-cells. Cell Immunol. 197, 83–90.

Kisseleva, T., Bhattacharya, S., Braunstein, J., Schindler, C.W., 2002. Signaling through the JAK-STAT pathway, recent advances and future challenges. Gene 285, 1–24.

Kubbies, M., 1992. Alteration of cell cycle kinetics by reducing agents in human peripheral blood lymphocytes from adult and senescent donors. Cell Prolif. 25, 157–168.

Lahdenpohja, N., Hurme, M., 1996. Naive (CD45RA+) T lymphocytes are more sensitive to oxidative stress-induced signals than memory (CD45RO+) cells. Cell Immunol. 173, 282–286.

Langlet, C., Bernard, A.-M., Drevo, P., He, H.-T., 2000. Membrane rafts and signaling by the multichain immune recognition receptors. Curr. Opin. Immunol. 12, 250–255.

Li, M., Walter, R., Torres, C., Sierra, F., 2000. Impaired signal transduction in mitogen activated rat splenic lymphocytes during aging. Mech. Ageing Dev. 113, 85–99.

Liu, B., Carle, K.W., Whisler, R.L., 1997. Reductions in the activation of ERK and JNK are associated with decreased IL-2 production in T cells from elderly humans stimulated by the TCR/CD3 complex and costimulatory signals. Cell Immunol. 182, 79–88.

Lohmiller, J.J., Roellich, K.M., Toledano, A., Rabinovitch, P.S., Wolf, N.S., Grossmann, A., 1996. Aged murine T-lymphocytes are more resistant to oxidative damage due to the predominance of the cells possessing the memory phenotype. J. Gerontol. A. Biol. Sci. Med. Sci. 51, B132–B140.

Luqman, M., Bottomly, K., 1992. Activation requirements for CD4+ T cells differing in CD45R expression. J. Immunol. 149, 2300–2306.

Marais, R., Wynne, J., Treisman, R., 1993. The SRF accessory protein Elk-1 contains a growth factor-regulated transcriptional activation domain. Cell 73, 381–393.

Matthews, R.P., Guthrie, C.R., Wailes, L.M., Zhao, X., Means, A.R., McKnight, G.S., 1994. Calcium/calmodulin-dependent protein kinase types II and IV differentially regulate CREB-dependent gene expression. Mol. Cell. Biol. 14, 6107–6116.

Miranti, C.K., Ginty, D.D., Huang, G., Chatila, T., Greenberg, M.E., 1995. Calcium activates serum response factor-dependent transcription by a Ras- and Elk-1-independent mechanism that involves $Ca2+$/calmodulin-dependent kinase. Mol. Cell. Biol. 15, 3672–3684.

Mustelin, T., Coggshall, K.M., Altman, A., 1989. Rapid activation of the T-cell tyrosine protein kinase p56lck by the CD45 phosphotyrosine phosphatase. Proc. Natl. Acad. Sci. 86, 6302–6306.

Myung, P.S., Boerthe, N.J., Koretzky, G.A., 2000. Adapter proteins in lymphocyte antigen-receptor signaling. Curr. Opin. Immunol. 12, 256–266.

Nada, S., Yagi, T., Takeda, H., Tokunaga, T., Nakagawa, H., Ikawa, Y., Okada, N., Aizawa, S., 1993. Constitutive activation of Src family kinases in mouse embryos that lack Csk. Cell 73, 1125–1135.

Negoro, S., Hara, H., Miyata, S., Saiki, O., Tanaka, T., Yoshizaki, K., Igarashi, T., Kishimoto, S., 1986. Mechanism of age related decline in antigen-specific T cell proliferative response: IL-2 receptor expression and recombinant IL-2 induced proliferative response of purified TAC-positive T cells. Mech. Ageing Dev. 36, 223–241.

Pahlavani, M.A., Harris, M.D., 1998. Effect of in vitro generation of oxygen free radicals on T cell function in young and old rats. Free Radic. Biol. Med. 25, 903–913.

Pahlavani, M.A., Vargas, D.M., 1999. Age-related decline in activation of calcium/calmodulin-dependent phosphatase calcineurin and kinase CaMK-IV in rat T cells. Mech. Ageing Dev. 112, 59–74.

Pahlavani, M.A., Vargas, D.M., 2000. Influence of aging and caloric restriction on activation of Ras/MAPK, calcineurin, and CaMK-IV activities in rat T cells. Proc. Soc. Exp. Biol. Med. 223, 163–169.

Pahlavani, M.A., Harris, M.D., Richardson, A., 1998. Activation of p21ras/MAPK signal transduction molecules decreases with age in mitogen-stimulated T cells from rats. Cell Immunol. 185, 39–48.

Patel, H.R., Miller, R.A., 1992. Age-associated changes in mitogen-induced protein phosphorylation in murine T lymphocytes. Eur. J. Immunol. 22, 253–260.

Pawelec, G., Barnett, Y., Forsey, R., Frasca, D., Globerson, A., McLeod, J., Caruso, C., Franceschi, C., Fulop, T., Gupta, S., Mariani, E., Mocchegiani, E., Solana, R., 2002. T cells and ageing, January 2002 update. Front. Biosci. 7, d1056–d1183.

Pulverer, B.J., Kyriakis, J.M., Avruch, J., Nikolakaki, E., Woodgett, J.R., 1991. Phosphorylation of c-jun mediated by MAP kinases. Nature 353, 670–674.

Quadri, R.A., Plastre, O., Phelouzat, M. A., Arbogast, A , Proust J J 1996. Age-related tyrosine specific protein phosphorylation defect in human T lymphocytes activated through CD3, CD4, CD8 or the IL-2 receptor. Mech. Ageing Dev. 88, 125–138.

Ravichandran, K.S., Lee, K.K., Songyang, Z., Cantley, L.C., Burn, P., Burakoff, S.J., 1993. Interaction of Shc with the zeta chain of the T cell receptor upon T cell activation. Science 262, 902–905.

Rincon, M., Flavell, R.A., Davis, R.A., 2000. The JNK and p38 MAP kinase signaling pathways in T cell-mediated immune responses. Free Radic. Biol. Med. 28, 1328–1337.

Sharma, H.K., Prasanna, H.R., Rothstein, M., 1980. Altered phosphoglycerate kinase in aging rats. J. Biol. Chem. 255, 5043–5050.

Shi, J., Miller, R.A., 1992. Tyrosine-specific protein phosphorylation in response to anti-CD3 antibody is diminished in old mice. J. Gerontol. 47, B147–B153.

Shi, J., Miller, R.A., 1993. Differential tyrosine-specific protein phosphorylation in mouse T lymphocyte subsets. Effect of age. J. Immunol. 151, 730–739.

Tamir, A., Eisenbraun, M.F., Garcia, G.G., Miller, R.A., 2000. Age-dependent alterations in the assembly of signal transduction complexes at the site of T cell/APC interaction. J. Immunol. 165, 1243–1251.

Tinkle, C.W., Lipschitz, D., Ponnappan, U., 1998. Decreased association of p56lck with CD4 may account for lowered tyrosine kinase activity in mitogen-activated human T lymphocytes during aging. Cell Immunol. 186, 154–160.

Utsuyama, M., Wakikawa, A., Tamura, T., Nariuchi, H., Hirokawa, K., 1997. Impairment of signal transduction in T cells from old mice. Mech. Ageing Dev. 93, 131–144.

van Oers, N.S., Weiss, A., 1995. The syk/ZAP-70 protein tyrosine kinase connection to antigen receptor signaling processes. Semin. Immunol. 7, 227–236.

Wang, J., Eck, M.J., 2003. Assembling atomic resolution views of the immunological synapse. Curr. Opin. Immunol. 15, 286–293.

Wegner, M., Cao, Z., Rosenfeld, M.G., 1992. Calcium-regulated phosphorylation within the leucine zipper of C/EBP beta. Science 256, 370–373.

Weiss, A., 1993. T cell antigen receptor signal transduction: a tale of tails and cytoplasmic protein-tyrosine kinases. Cell 73, 209–212.

Whisler, R.L., Bagenstose, S.E., Newhouse, Y.G., Carle, K.W., 1997. Expression and catalytic activities of protein tyrosine kinases (PTKs) Fyn and Lck in peripheral blood T cells from elderly humans stimulated through T cell receptor (TCR)/CD3 complex. Mech. Ageing Dev. 98, 57–73.

Whisler, R.L., Chen, M., Liu, B., Newhouse, Y.G., 1999. Age-related impairments in TCR/CD3 activation of ZAP-70 are associated with reduced tyrosine phosphorylation of zeta-chains and p59fyn p56lck in human T cells. Mech. Ageing Dev. 111, 49–66.

Whisler, R.L., Karanfilov, C.I., Newhouse, Y.G., Fox, C.C., Lakshmann, R.R., Liu, B., 1998. Phosphorylation and coupling of zeta-chains to activated T-cell receptor (TCR)/CD3 complexes from peripheral blood T-cells of elderly humans. Mech. Ageing Dev. 105, 115–135.

Whisler, R.L., Newhouse, Y.G., Bagenstose, S.E., 1996. Age-related reductions in the activation of mitogen-activated protein kinases p44mapk/ERK1 and p42mapk/ERK2 in human T cells stimulated via ligation of the T cell receptor complex. Cell Immunol. 168, 201–210.

Whitehurst, C.E., Boulton, T.G., Cobb, M.H., Geppert, T.D., 1992. Extracellular signal-regulated kinases in T cells: anti-CD3 and 4 beta-phorbol 12-myristate 13-acetate-induced phosphorylation and activation. J. Immunol. 148, 3230–3237.

Yang, D., Miller, R.A., 1999. Cluster formation by protein kinase c theta during murine T cell activation. Cell Immunol. 195, 28–36.

**Advances in
Cell Aging and
Gerontology**

Cyclic nucleotide signaling in vascular and cavernous smooth muscle: aging-related changes

Ching-Shwun Lin* and Tom F. Lue

*Department of Urology, Knuppe Molecular Urology Laboratory, University of California,
San Francisco, CA 94143-1695, USA.
*Tel.: + 1-415-353-7205; fax: + 1-415-353-9586.
E-mail address: clin@urol.ucsf.edu*

Contents

1. Introduction

Vascular diseases are responsible for one-third of deaths worldwide and represent the most important cause of death in industrialized nations (WHO, 2002). In the USA, more than 40% of deaths result from vascular diseases, especially

Advances in Cell Aging and Gerontology, vol. 16, 57–106
DOI: 10.1016/S1566-3124(04)16004-5

coronary heart disease (CHD) and cerebrovascular diseases (CVD) (Anderson, 2002; Arias and Smith, 2003). While atherosclerotic coronary occlusion is the most important cause of premature death (especially among men) in industrialized societies, its main impact falls on the elderly. In fact, more than 80% of those who die from coronary heart disease are older than 65 (Shepherd, 2001). Similarly, CVD, which are the third leading cause of death in the USA, affect primarily the elderly. Alzheimer's disease (AD), which shares many characteristics with vascular dementia (de la Torre, 2002; Pansari et al., 2002), is the eighth leading cause of death in the USA (Anderson, 2002) and again it affects mainly the elderly. Erectile dysfunction (ED), which impacts the quality of life of more than 20 million American men and their spouses, is another socioeconomically important vascular disorder that afflicts mainly aging men (Lue, 2000). Together, these four vascular disorders (CHD, CVD, AD, and ED) weigh enormously on our humanity especially as the elderly segment of the population is rapidly expanding.

One common denominator of these aging-related vascular disorders is the inability of the affected vascular (or, in the case of ED, cavernous) smooth muscle to relax properly. As smooth muscle tone is principally regulated through cyclic nucleotide signaling, it is conceivable that aging might adversely affect certain elements or processes of this pathway, thereby predisposing the individual to hypertension. Furthermore, the normally quiescent vascular smooth muscle cells (VSMCs) often become proliferative in neointima formation as the hypertensive vessels become atherosclerotic (Ross, 1997; Newby and Zaltsman, 2000). There is abundant evidence that activation of the proliferative phenotype in VSMCs is also regulated by cyclic nucleotide signaling (Koyama et al., 2001). As such, cyclic nucleotide signaling, by virtue of its regulation of smooth muscle tone as well as cellular proliferation, plays key roles in the pathogenesis of many prominent diseases.

Briefly, cyclic nucleotide signaling in the VSM starts with the release of an extracellular signaling molecule from the endothelium or nerve endings. The signaling molecule then diffuses inside VSMC or binds to a receptor in the plasma membrane of VSMC (Fig. 1). The signaling molecule itself or the ligand-bound receptor activates adenyl cyclase (AC) or guanyl cyclase (GC), which then catalyzes the conversion of ATP to cAMP or GTP to cGMP. Subsequent binding of cAMP to protein kinase A (PKA), or binding of cGMP to protein kinase G (PKG), results in the activation of the respective kinase, which then phosphorylates specific downstream targets such as enzymes, transcription factors, and ion channels. The outcomes are changes in cell metabolism, motility, polarity, and/or gene expression. Termination of the signaling event is carried out by phospho-diesterases (PDEs) that catalyze the hydrolysis of cAMP to AMP and/or cGMP to GMP.

While commonly regarded as a urinary organ, the penis, in serving its repro-ductive function, is in fact a vascular organ. Structurally, the penis, like all other body organs, has a "regular" vascular system consisting of incoming arteries and outgoing veins. Uniquely however, each of the two bodies of erectile tissue, or the corpora cavernosa, is composed of a conglomeration of sinusoids that are lined by

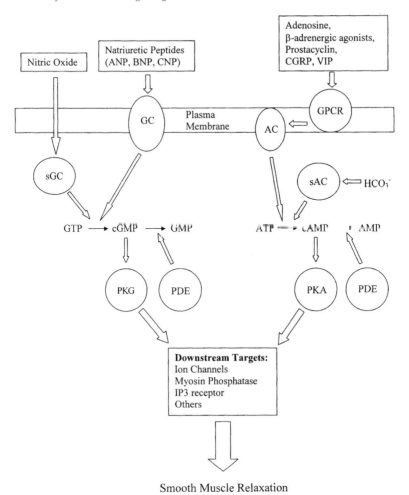

Fig. 1. Cyclic nucleotide signaling cascades leading to vascular smooth muscle relaxation. Block arrows indicate binding and/or activation. Binding of HCO_3^- to sAC has not been demonstrated in vascular smooth muscle cells.

endothelial cells (Fig. 2). The sinusoids are relatively small when the penis is in the flaccid state; however, they become enlarged rapidly as arterial blood rushes in during erection. The entry of blood is controlled by cavernous smooth muscle cells (CSMCs) that form irregular layers around each sinusoid. Relaxation of CSMCs allows the entry of blood into the sinusoids and is initiated by nitric oxide (NO) that is released from the cavernous nerves (which become dorsal nerves, Fig. 2) of a sexually stimulated individual. NO binds to and activates GC, GC catalyzes the synthesis of cGMP, cGMP activates PKG, and PKG phosphorylates downstream protein targets, resulting in the relaxation of CSMCs.

Like other signal transduction pathways (e.g., those of growth factors and cytokines), cyclic nucleotide signaling operates not in a linear pathway as depicted

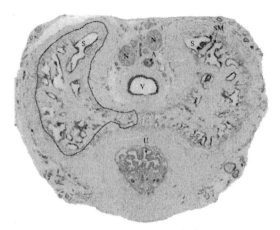

Fig. 2. Cross-section of rat penis. The tissue section is stained for PKG-I (originally stained brown), which is expressed in all smooth muscles including cavernous smooth muscle (SM), dorsal artery (A), and dorsal vein (V). For ease of illustration, the boundary of one of the corpora cavernosa is traced. The urethra (U) is also indicated. Within each corpus cavernosum, numerous sinusoids (S) are each lined with a single layer of endothelial cells (obscured by SM). Nerve fibers branching from the dorsal nerves (N) transmit signaling molecules such as NO, CGRP, and VIP to SM. Original magnification is 25×.

above, but rather in a complex network that encompasses many interweaving pathways. However, because the linear pathway is the best characterized and appears to be the most important in the physiological as well as pathological processes, this chapter's discussion of cyclic nucleotide signaling will be organized along this pathway. Whenever relevant and available, any branching pathways will be drawn at the appropriate junctions along this major pathway. It should be pointed out that the cAMP- and cGMP-signaling pathways, despite some degree of intermingling, are practically two separate entities and therefore will be discussed as such. It should also be forewarned that blood vessels are known to possess tissue/organ-specific properties; therefore, the signaling pathways discussed in this chapter are not necessarily applicable to all types of vessels. Finally, because very little is known about age-related changes in the vascular system and much less when it comes to cyclic nucleotide signaling, this chapter covers mostly general issues of vascular cyclic nucleotide signaling. By providing information on what has been done and what has not, it is our hope that interested readers will be able to locate areas that deserve better attention and pursue them with future research.

2. The signals

VSMCs are situated in the medial layer (the media) of blood vessels and they receive extracellular signals from the endothelium in the intima or from nerve fibers along the media-adventitia border. In response to these signals VSMCs relax or contract as they perform their physiological duties in their normally nonproliferative state. When vessels are injured primarily in atherosclerotic lesions or secondarily following balloon angioplasty, VSMCs in the injured area undergo a transition from the contractile phenotype to the synthetic phenotype in which cellular proliferation becomes VSMC's principal function (Campbell and Campbell, 1990; Newby and Zaltsman, 2000). Both the contractile and the synthetic phenotypes are intimately regulated by cyclic nucleotide signaling that is initiated by a large number of signaling molecules including nitric oxide, acetylcholine, prostacyclin, etc. Each of these signaling molecules stimulates the production of cAMP or cGMP by direct or indirect activation of adenyl or guanyl cyclase, as depicted in Fig. 1 and as explained below.

1. Direct activation of the soluble guanyl cyclase (sGC) in the cytoplasm, as exemplified by NO. After its release from the endothelium, NO enters VSMCs by diffusion. It then binds to and activates sGC that catalyzes the synthesis of cGMP; thus, sGC acts as both a receptor and an effector for NO
2. Direct activation of the particulate guanyl cyclase (pGC) in the cytoplasmic membrane, as exemplified by natriuretic peptides. These peptides are secreted from heart, brain, and/or the endothelium and act as endocrine or paracrine hormones on blood vessels. They bind to and activate either guanyl cyclase-A (GC-A) or -B (GC-B) located in the cytoplasmic membrane of VSMCs. Thus, GC-A and GC-B serve as both receptors and effectors for natriuretic peptides.
3. Indirect activation of the particulate adenyl cyclase (pAC) in the cytoplasmic membrane via direct activation of a G-protein coupled receptor (GPCR). The vast majority of vasoregulators act through this route. Each of these signaling molecules binds to a specific receptor that belongs to the GPCR superfamily, causing the receptor to augment the exchange of GTP for GDP on the α subunit of the heterotrimeric G protein. Binding of GTP to $G\alpha$ subunit causes it to dissociate from the $G\beta\gamma$ subunits, and both the $G\alpha$ subunit and the $G\beta\gamma$ subunits are now able to activate pAC, triggering the production of cAMP. Because the $G\alpha$ subunit has an intrinsic GTPase activity that hydrolyzes GTP to GDP, it can convert itself from the GTP-bound form to the GDP-bound form. The GDP-bound, inactive $G\alpha$ then reassociates with the $G\beta\gamma$ subunits, thus returning the G protein to the resting state and terminating the transmission of the initial signal. While this signaling pathway is used by most cyclic nucleotide-signaling molecules, there are instances when AC can be inhibited rather than stimulated (Patel et al., 2001). For example, the $\alpha2$ adrenoceptors are predominately coupled to the G protein containing an inhibitory (α_i) rather than a stimulatory (α_s) α subunit. Therefore, their activation by adrenaline and adrenaline agonists results in the inhibition of AC (Guimaraes and Moura, 2001).

One additional signaling pathway that has not been specifically described in the vascular system is the direct activation of the soluble adenyl cyclase (sAC) by bicarbonate ion (HCO_3^-). Unlike other signaling molecules discussed above, HCO_3^- is produced and signals intracellularly. It binds to and activates sAC, which thus acts as both the receptor and effector for HCO_3^-. Because cellular levels of HCO_3^- reflect the output of the energy-generating metabolic processes, the HCO_3^-–sAC pathway is thought to represent a metabolic feedback mechanism (Zippin et al., 2001). Such a mechanism is believed to operate in various cell types (Chen et al., 2000b). However, its existence in VSMCs has not been reported.

In the following subsections some of the best-characterized molecules that signal through the cyclic nucleotide-signaling pathways will be discussed. Those that signal primarily through the cAMP pathway include adenosine, adrenergic hormones, calcitonin gene-related peptide (CGRP), prostaglandins, and vasoactive intestinal peptide (VIP). Those that signal primarily through the cGMP pathway are natriuretic peptides and NO.

2.1. Adenosine

Adenosine is a natural cardioprotective agent produced by the heart during episodes of ischemia and hypoxia (Clemens and Forrester, 1981) and in response to increases in oxygen demand during exercise (Duncker et al., 1995). It is now recognized that adenosine can be released from a variety of cells as a result of increased metabolic rates and its actions on the vasculature are most prominent when oxygen demand is high (Tabrizchi and Bedi, 2001). However, the vascular response to the action of adenosine can be either relaxation or constriction, depending on which type of the adenosine receptor (AR) is activated.

Currently, four AR subtypes (A_1, A_{2A}, A_{2B}, and A_3) belonging to the GPCR superfamily have been recognized (Olah and Stiles, 2000; Fredholm et al., 2001; Tabrizchi and Bedi, 2001; Klinger et al., 2002). In general, the A_1 receptor is believed to be coupled to G_i and G_o proteins of the G protein family, and its activation results in inhibition of adenyl cyclase (AC) and activation of phospholipase C, both of which lead to vasoconstriction. The A_2 receptors are coupled to the G_s proteins (Marala and Mustafa, 1993), and their activation results in the stimulation of AC and thus vasorelaxation. The A_3 receptor is coupled to G_i and G_q proteins, and its activation results in the activation of phospholipase C/D and the inhibition of AC, leading to vasoconstriction. The differential distribution of these AR subtypes largely determines whether a particular vessel relaxes or contracts as a result of adenosine stimulation (Tabrizchi and Bedi, 2001).

Age-related decline in dilation responses to adenosine has been shown in rat aorta (Moritoki et al., 1990; Hinschen et al., 2001), pial arterioles (Jiang et al., 1992b), and coronary arteries (Moritoki et al., 1990; Hinschen et al., 2001, 2003). In the case of coronary arteries, the age-related reduction of response to adenosine does not involve A_1 or A_{2A} but may involve A_{2B} receptor (Hinschen et al., 2003). No significant age-related difference was found in the dilation response to adenosine in the dorsal hand vein of human subjects (Ford et al., 1992).

Whether adenosine plays a role in physiological penile erection is unknown (Simonsen et al., 2002). Intracavernous injection of adenosine has been shown to cause full erection in the dog (Takahashi et al., 1992), and the A_{2A} receptor was shown to mediate the relaxation effects of adenosine in human and rabbit corpus cavernosum (Mantelli et al., 1995; Filippi et al., 2000). However, intracavernous injection of adenosine failed to induce erection in human volunteers, possibly due to simultaneous relaxation of penile veins (Filippi et al., 2000). This is in contrast with prostaglandin E1 (see Section 2.4), which relaxes cavernous smooth muscle but contracts penile veins, thus allowing erection (Filippi et al., 2000).

2.2. Adrenergic hormones

Sympathetic nerves coursing along the media-adventitia border of blood vessels release many vasoactive hormones including adrenaline and noradrenaline, which act on the α-type adrenoceptors in VSMCs, causing vasoconstriction. In rare instances, adrenaline and noradrenaline can produce vasodilation by acting on the β-type adrenoceptors (Begonha et al., 1995; Guimaraes and Moura, 2001). Selective α- and β-type adrenergic antagonists (alpha- and beta-blockers) have long been used as therapeutic agents to modulate vascular tone (Freis, 1997; Reid, 1999). Particularly, the beta-blockers that are supposed to increase peripheral vascular resistance have been used as "first-line" treatment for hypertension. The mechanisms for the hypotensive effects of beta-blockers are not understood and the benefits of beta-blockers in treating the elderly have been questioned (Beevers, 1998; Grossman and Messerli, 2002).

The α-type adrenoceptors include α1A, α1B, α1D, α2A/D, α2B, and α2AC. The β-type adrenoceptors include β1, β2, and β3. All nine adrenoceptor subtypes have been identified in VSMCs, and the degree of involvement of each subtype depends on the vascular bed and the species (Guimaraes and Moura, 2001). While these receptors all belong to the GPCR family, they produce different effects depending on the type of G protein to which each adrenoceptor subtype is coupled. It is well known that the α1-adrenoceptors are mainly coupled to the Gq/11 protein that stimulates phospholipase C rather than AC (Zhong and Minneman, 1999; Garcia-Sainz et al., 2000), and the α2-adrenoceptors are predominantly coupled to the inhibitory G protein that inhibits the activity of AC (Wise et al., 1997). As such, the α-type adrenoceptors are not directly involved in cyclic nucleotide signaling.

The β-adrenoceptors are coupled to the stimulatory G protein that stimulates the activity of AC. These β-adrenoceptors are therefore among the initiators of cAMP signaling that produces vasodilation. The β2-adrenoceptors represent the predominant subtype in most vascular smooth muscles, although β1- and β3-adrenoceptors may also contribute to vasodilation. It is well known that β-adrenoceptor-mediated vascular smooth muscle relaxation decreases with increasing age (Volicer et al., 1983; Pan et al., 1986; Tsujimoto et al., 1986; Docherty, 1990; Crass et al., 1992; Ford et al., 1992; Dohi et al., 1995; Harada et al., 1996; Guimaraes and Moura, 1999; Werstiuk and Lee, 2000). However, no difference in the expression levels of β-adrenoceptors in various vascular beds has been found between young and old humans or animals (Tsujimoto et al., 1986; Mader, 1992; Guimaraes and Moura, 1999, 2001). Furthermore, no age-related difference in the expression level of inhibitory or stimulatory G protein was found (Chin et al., 1996a).

As such, the mechanisms responsible for the reduced β-adrenergic responses in older individuals occur downstream in the signaling pathway.

Both α1- and α2-adrenoceptors have been demonstrated in erectile tissues with the former being predominant in cavernous smooth muscle and the latter being more abundant on the arterioles supplying the cavernous spaces (Hedlund and Andersson, 1985a; Saenz de Tejada et al., 1989). α-Adrenergic stimulation results in contraction of cavernous smooth muscle and arteriolar vessels, leading to detumescence. Intracavernous injection of phentolamine, an α-adrenergic antagonist, does not produce rigid erection when used alone, but can improve success rates when used in combination with papaverine, which inhibits phosphodiesterases (see Section 12) (Stief and Wetterauer, 1988; Lue, 2000). β-Adrenergic agonists have been shown to relax isolated human corporal and arterial smooth muscles (Hedlund and Andersson, 1985a). However, being outnumbered by α-receptors by a factor of 10 to 1, β-receptors probably play a limited role in erectile function (Levin and Wein, 1980).

2.3. Calcitonin gene-related peptide family

Calcitonin gene-related peptide (CGRP), amylin, and adrenomedullin are members of the CGRP family. These short-chain peptides are potent vasodilators released from perivascular nerve fibers (Brain et al., 1985; Kawasaki et al., 1988; Ishiyama et al., 1993; Bell and McDermott, 1996; van Rossum et al., 1997; Wimalawansa, 1997; Zhu et al., 1997; Juaneda et al., 2000; Minamino et al., 2000). They act through the calcitonin receptor-like receptor (CRLR) that belongs to the GPCR superfamily (Conner et al., 2002). However, unlike other GPCR family members, the CRLR itself is nonfunctional unless aided by two additional proteins, namely, the CGRP-receptor component protein (RCP) and the receptor activity modifying protein (RAMP1, RAMP2, or RAMP3). RCP interacts with CRLR, RAMP1, and RAMP2 and facilitates signal transduction by adrenomedullin and CGRP (Prado et al., 2001). The RAMP proteins act as molecular chaperones for trafficking of CRLR to the cell surface. They also influence CRLR's preferential binding to either CGRP or adrenomedullin – RAMP1 favors CGRP, and RAMP2 or RAMP3 favors adrenomedullin (McLatchie et al., 1998; Fraser et al., 1999).

CGRP-induced vasodilating responses in different vascular beds are mediated via different cell-signaling pathways (Marshall, 1992). CGRP dilates most vessels through the cAMP-signaling pathway independent of the endothelium; however, its vasodilating effects in some vessels are dependent on the endothelium and are mediated by NO and cGMP (Grace et al., 1987; Gray and Marshall, 1992; Hayakawa et al., 1999). Similarly, the vasorelaxant effects of adrenomedullin are mediated through both cAMP-dependent and -independent pathways (Minamino et al., 2000). Additionally, adrenomedullin inhibits the proliferation of VSMCs via the cAMP signaling pathway (Kano et al., 1996), but stimulates the proliferation of quiescent VSMCs via the p42/p44 ERK/MAP kinase pathway (Iwasaki et al., 1998). Intravenous injection of adrenomedullin in rats results in long-lasting hypotension (Ishiyama et al., 1993). Systemic and local delivery of adreno-medullin gene attenuates hypertension, myocardial infarction, renal injury, and cardiovascular remodeling in animal models via cAMP- and cGMP-signaling pathways (Chao et al., 2001). Intratracheal administration of CGRP gene also increases lung cAMP levels and attenuates pulmonary hypertensive disorders (Champion et al., 2000).

Age-related changes in CGRP content have been observed in rat tissues, including a decline in arteries and increase in veins (Wimalawansa, 1992). In the aorta of old rats (> 1 year), CGRP-containing nerve fibers are no longer detectable (Connat et al., 2001). The CGRP levels found in the penis, bladder, kidney, testis, and adrenal gland gradually increased up to maturity, and then rapidly declined in the aging rats (Wimalawansa, 1992). When CGRP is administered intracaver-nously in patients suffering from age-related ED, a dose-related increase in penile arterial inflow (and erection) occurs (Stief et al., 1991; Djamilian et al., 1993; Truss et al., 1994). Adenovirus-mediated gene transfer of CGRP also enhances erectile responses in aged rats, apparently through an increase of cAMP levels in the corpora cavernosa (Bivalacqua et al., 2001). Intracavernous administration of adrenomedullin also results in cavernous relaxation; however, the effect is through a NO-cGMP instead of the cAMP pathway (Nishimatsu et al., 2001).

2.4. Prostaglandins

Prostaglandins (PGs) are a family of eicosanoids capable of initiating numerous biological functions including vasodilation (Funk, 2001). They are synthesized by most cells in our body and act as autocrine and paracrine mediators. Blood vessels synthesize principally PGI2 (prostacyclin), PGE2, PGD2, PGF2α, and thromboxane A2 (TXA2) (Jeremy et al., 1988). In most vessels PGI2, PGE2, and PGD2 act as vasodilators while PGF2 and TXA2 are vasoconstrictors. Inhibition of PG synthesis has been shown to increase basal vascular tone and to attenuate responses to vasodilating agents (Hadhazy et al., 1988). Therefore, the overall role of endogenous vascular PGs is vasodilatation. In particular, PGI2 is the most important vasodilator produced by the endothelium that signals through the cAMP pathway (Wise and Jones, 1996).

The prime mode of PG action is through specific PG receptors that all belong to the GPCR family. There are at least nine known PG receptor subtypes in mouse and man, as well as several additional splice variants with divergent carboxyl termini (Breyer et al., 2000; Narumiya and FitzGerald, 2001). Four of the subtypes (EP1–EP4) bind PGE2, two (DP1 and DP2) bind PGD2, and the other three subtypes (FP, IP, and TP) bind PGF2α, PGI2, and TxA2, respectively. On the basis of signaling attributes, the PG receptors are classified into three types. The "relaxant" receptors IP, DP1, EP2, and EP4 are coupled to an α_s-containing G protein and therefore capable of stimulating AC to increase intracellular cAMP. The "contractile" receptors EP1, FP, and TP are coupled to an α_q-containing G protein, which activates phospholipase C instead of AC. These contractile receptors therefore do not signal through the cAMP pathway and their signaling outcome is an increase of intracellular calcium. The EP3 receptor is also a contractile receptor, but it is coupled to an α_i-containing G protein that inhibits AC to result in a decrease of cAMP formation.

Animal and human corpus cavernosum also produces several PGs including PGF2α, PGE2, PGI2, and TXA2 (Saenz de Tejada et al., 1988). In studies on isolated human penile tissue, different PGs have been shown to elicit different effects on human corpus cavernosum, corpus spongiosum, and cavernous artery (Hedlund and Andersson, 1985b). While PGF2α, PGI2, and TXA2 contract the corpus cavernosum and corpus spongiosum, PGE2 (but not PGI2) relaxes the corpus cavernosum and spongiosum that have been precontracted with noradrenaline or PGF2α. Likewise, PGE2, but not PGI2, increases the arterial blood flow when injected into monkey's corpus cavernosum (Aboseif et al., 1989; Bosch et al., 1989). Therefore, while PGI2 is the predominant vasorelaxant in blood vessels, its action in the erectile tissue is either contractile or neutral. This disparity in the action of PGI2 between blood vessels and the erectile tissue and the difference between the activities of PGI2 and PGE2 in the erectile tissue are most likely due to differences in the distribution of PG receptors. Indeed, a recent study showed that in the corpus cavernosum the relaxant effects of prostanoids are mediated by EP2- and/or EP4-receptors (for PGE2), but not IP receptor (for PGI2) (Angulo et al., 2002).

2.5. Vasoactive intestinal peptide

Many systemic blood vessels are innervated by nerve fibers that release vasoactive intestinal peptide (VIP), causing vascular smooth muscle relaxation (Larsson et al., 1976; Dauphin and MacKenzie, 1995; Zhu et al., 1997; Henning and Sawmiller, 2001). On a molar basis, VIP is 50–100 times more potent than acetylcholine as a vasodilator (Fahrenkrug, 1989). In humans and animals, VIP increases coronary blood flow significantly when administered into the coronary artery or intravenously (Brum et al., 1986; Itoh et al., 1990). The coronary dilator effect of VIP, at maximal doses, is significantly greater than that of β-adrenergic agonist isoproterenol (Anderson et al., 1988).

The mature VIP peptide contains 28 amino acid residues and is derived from prepro-VIP, which consists of 170 amino acid residues. The primary structure of VIP is closely related to pituitary adenyl cyclase-activating polypeptide (PACAP) and, to a much lesser extent, to secretin, glucagon, gastric inhibitor polypeptide, and helodermin-like peptides (Klimaschewski, 1997). Two subtypes of VIP receptors, VPAC1 and VPAC2, belonging to the GPCR family have been cloned from human and rat tissues. VPAC1 is widely distributed in many tissues including various blood vessels (Henning and Sawmiller, 2001). Expression of VPAC2 in VSMCs has also been reported (Busto et al., 2000; Reubi, 2000; Schmidt et al., 2001). Through these receptors VIP activates AC in a dose-dependent manner in a wide variety of blood vessels (Henning and Sawmiller, 2001). While the expression of the cognate receptors in VSMCs allows VIP to exert its vasorelaxant effect through direct activation of the cAMP pathway, in certain vessels the vasorelaxant effect of

VIP depends on an intact endothelium (Henning and Sawmiller, 2001). This endothelium-dependent effect is mediated by activation of lipoxygenase in the rat aorta and by NO activation of GC in the human uterine artery (Davies and Williams, 1984; Ignarro et al., 1987b; Jovanovic et al., 1998).

Connat et al. (2001) used immunofluorescence technique to examine age-related changes in peptidergic innervation in the aorta. VIP-positive fibers could be detected in young rats (1–6 months of age) but not in old rats (1 year or older). Age-related reduction of VIP expression or innervation has been seen in many other tissues such as the gastrointestinal tract (El-Salhy and Sandstrom, 1999), hypothalamus (Zhou and Swaab, 1999), sweat gland (Vilches et al., 2002), and hippocampus (Vela et al., 2003). Of clinical importance is that older individuals (> 60 years) are less able than younger patients to activate VIP expression after suffering acute myocardial infarction (AMI) (Lucia et al., 1996). These older individuals and younger ones who have low plasma VIP concentrations are less likely to survive from AMI (Lucia et al., 1996).

The human and animal penis is richly supplied with nerves containing VIP (Andersson and Wagner, 1995; Andersson, 2001). VPAC2, but not VPAC1 mRNA, has also been identified in CSMCs (Guidone et al., 2002). However, it has been shown that VIP release is not essential for neurogenic relaxation of human cavernous smooth muscle (Pickard et al., 1993) and VIP failed to stimulate cAMP production in cultured human CSMCs (Palmer et al., 1994). *In vitro* and *in vivo* studies have not been able to settle whether VIP has a role in cavernous smooth muscle relaxation (Andersson, 2001). Aging does not affect the expression level of VIP mRNA in rat cavernous smooth muscle (Shen et al., 2000). Intracavernous injection of VIP does not produce a rigid erection (Kiely et al., 1989) but improves success rates when combined with α-adrenergic antagonist, phentolamine (Dinsmore and Alderdice, 1998).

2.6. Natriuretic peptides

The natriuretic peptide family is involved in the regulation of cardiovascular homeostasis and consists of atrial natriuretic peptide (ANP), brain natriuretic peptide (BNP), and C-type natriuretic peptide (CNP) (Barr et al., 1996; Chen and Burnett, 1998; Matsuo, 2001; Rubattu and Volpe, 2001; Silberbach and Roberts, 2001; Suzuki et al., 2001). The mature ANP has 28 amino acid residues after cleavage from a precursor peptide (Atlas et al., 1984; Inagami et al., 1984; Kangawa and Matsuo, 1984). BNP and CNP have 32 and 22 amino acid residues, respectively, and are also cleavage products of precursor polypeptides (Matsuo, 2001; Rubattu and Volpe, 2001; Suzuki et al., 2001). Both ANP and BNP act mainly as cardiac hormones and are produced predominantly by the atrium and ventricle, respectively. Expression of ANP and BNP is greatly augmented in patients with congestive heart failure and in animal models of ventricular hypertrophy or cardiomyopathy (Saito et al., 1987; Yasue et al., 1989; Takemura et al., 1991; Di Nardo et al., 1993; Hasegawa et al., 1993; Chen and Burnett, 2000;). In contrast, CNP does not act as a cardiac hormone but as a neuropeptide or an endothelium-derived autocrine/paracrine regulator. In particular, while most other vasorelaxants are biased toward arterial relaxation, CNP is predominantly a venodilator (Barr et al., 1996; Chen and Burnett, 1998). The endothelial production of CNP is remarkably augmented by various cytokines and growth factors such as transforming growth factor-β and tumor necrosis factor-α, suggesting the pathophysiological significance of CNP in the process of various vascular disorders (Ogawa et al., 1995).

Whereas ANP and BNP are ligands for the NPR-A receptor, CNP is a ligand for the NPR-B receptor. NPR-A and NPR-B receptors are members of the GC family and are therefore also called GC-A and GC-B. ANP, BNP, and CNP also bind to a common receptor, NPR-C, that is a truncated form of NPR-A. NPR-C lacks guanyl cyclase activity (Lowe et al., 1990; Potter and Hunter, 2001) and should not be confused with GC-C of the GC family (see Section 4). In many tissues, NPR-C is more abundant than NPR-A or NPR-B, and due to its lack of GC activity, NPR-C has been regarded a "clearance" receptor whose primary function is removal of natriuretic peptides from the system (Lowe et al., 1990; Lucas et al., 2000; Potter and Hunter, 2001). This assumption is now proven inaccurate, as explained below.

While prostacyclin and NO are the best-known endothelium-derived factors that relax vascular smooth muscle by signaling through the cAMP and cGMP pathways, respectively, a third endothelium-derived factor that mediates the vasorelaxation effects of acetylcholine and bradykinin independent of NO or prostacyclin has been recognized (Triggle and Ding, 2002). Because of its association with endothelium-dependent hyperpolarization of VSMCs, this factor has been called endothelium-derived hyperpolarizing factor or EDHF (Triggle and Ding, 2002). A recent study provides evidence that CNP is EDHF and it relaxes blood vessels by activating NPR-C but not NPR-B (Chauhan et al., 2003). Coupling of the activated NPR-C to G_i protein and the smooth muscle G protein-gated inwardly rectifying K^+ channel accounts for the vasorelaxing effect of CNP (Chauhan et al., 2003).

As a treatment for hypertension, direct infusion of ANP has limited success because of its short half-life in the circulation. To circumvent this limitation, adenovirus-mediated delivery of ANP gene has been attempted in animal models and the results showed amelioration of various cardiovascular disorders including hypertension, renal injury, and stroke (Lin et al., 1999). Adenovirus-mediated delivery of CNP has also been shown to markedly reduce neointimal formation in balloon-injured carotid and coronary arteries mainly through enhanced NO production (Ueno et al., 1997; Morishige et al., 2000; Qian et al., 2002; Yamahara et al., 2003).

The effects of ANP, BNP, and CNP on cGMP production and smooth muscle relaxation in isolated human and animal corpus cavernosum and in cultured CSMCs have been investigated (Kim et al., 1998; Guidone et al., 2002; Kuthe et al., 2003). The results indicate that CNP is the most potent natriuretic peptide and it relaxes the isolated cavernous smooth muscle by binding to NPR-B. However, whether CNP and NPR-B play a role in physiological erection remains to be investigated.

2.7. Nitric oxide

Nitric oxide was initially identified as a gaseous molecule released from various vasodilating compounds and capable of stimulating cGMP production via activation of GC in various tissues (Arnold et al., 1977). Later, NO was shown to be the EDRF (endothelium-derived relaxing factor) for blood vessels (Furchgott and Zawadzki, 1980; Ignarro et al., 1987a; Palmer et al., 1987; Murad, 1999). It is now known that NO is produced in many tissues in addition to endothelium and its vasorelaxing effect is mediated by cGMP. Because of its small size, NO can diffuse inside its target cell where it interacts with molecules containing iron in either a heme or iron–sulfur complex (Sato and Murota, 1995). The most physiologically relevant target for NO is sGC, and the NO–sGC–cGMP pathway is responsible for the vasorelaxing effect of many endothelium-dependent vasodilators, including histamine (Olesen et al., 1994), estrogens (White et al., 1995), insulin (Trovati et al., 1995), and corticotrophin-releasing hormone (Clifton et al., 1995), nitrovasodilators (Kelly and Smith, 1996), and acetylcholine (Iranami et al., 1996).

Agents that stimulate NO production have also been shown to attenuate VSMC proliferation, a key feature in neointimal formation after balloon angioplasty (Garg and Hassid, 1989). The inhibitory effect of NO is mediated through both cGMP-dependent and -independent pathways, the latter of which involves downregulation of cell cycle regulatory proteins (Tanner et al., 2000). NO may also prevent neointimal formation and pulmonary hypertension-associated vascular remodeling by inducing apoptosis of VSMCs via cGMP signaling (Pollman et al., 1996; Smith et al., 1998; Krick et al., 2001).

Synthesis of NO is catalyzed by nitric oxide synthases (NOS) that convert L-arginine and oxygen to L-citruline and NO. The three mammalian NOS isoforms are neuronal (nNOS), inducible (iNOS), and endothelial (eNOS) (Michel and Feron, 1997; Murad, 1999; Alderton et al., 2001; Davis et al., 2001; Hanafy et al., 2001). While nNOS and eNOS are preferentially expressed in neurons and endothelial cells, respectively, iNOS is expressed in most cell types with elevated expression when induced by stimuli such as bacterial lipopolysaccharide. Targeted deletion of the eNOS gene results in hypertension (Huang et al., 1995), increased neointimal growth in response to vascular injury (Moroi et al., 1998), and hyperplastic growth of vascular smooth muscle in response to arterial ligation (Rudic et al., 1998). All three NOS isoforms have been used in gene transfer studies to treat various cardiovascular diseases including cerebral spasm, atherosclerosis, restenosis, autograft vasculopathy, cardiac allograft diseases, and hypertension (Chen et al., 2002).

Reduction of NO bioavailability due to aging has been linked to impaired endothelium-dependent vasodilation in various human vascular beds (Zeiher et al., 1993; Taddei et al., 1995, 2001; Gerhard et al., 1996; Lyons et al., 1997; Andrawis et al., 2000; Woodman et al., 2002). The reduced NO bioavailability is caused by various mechanisms that are not fully understood but may include down-regulated NOS expression or activity (Maxwell, 2002). Cernadas et al. (1998) found that in the aorta of aging rats, the expression of both eNOS and iNOS isoforms was enhanced; however, the activity of eNOS was markedly reduced. The authors postulated that the concomitant elevation of iNOS expression and activity might have an inhibitory effect on the eNOS activity. However, several other studies appear to point to an association of aging with downregulated activity and expression of eNOS in the endothelium (Chou et al., 1998; Matsushita et al., 2001; Nakamura et al., 2002; Vaziri et al., 2002). The reason for the discrepancy between the study by Cernadas et al. and the other studies is unclear. It should be pointed out that the latter studies were conducted with genetically modified rats that are either spontaneously hypertensive or showing signs of accelerated aging.

While eNOS is principally responsible for the synthesis of NO in blood vessels, nNOS is responsible for the synthesis of NO that is released in the corpus cavernosum following sexual stimulation (Burnett et al., 1992). Downregulation of nNOS expression has been found in the corpus cavernosum of aging rats (Carrier et al., 1997; Dahiya et al., 1997) and NO-mediated corpus cavernous smooth muscle relaxation is impaired in aging rats (Cartledge et al., 2001). Despite its secondary role in erectile function, the endothelium-dependent cavernous smooth muscle relaxation is attenuated in the aging rabbit, and this aging-related defect is paradoxically accompanied with upregulated eNOS expression in both the endothelium and cavernous smooth muscle (Haas et al., 1998). Contradictory observation, however, was made in another study that showed downregulation of eNOS in the corpus cavernosum of aging rats (Rajasekaran et al., 2002).

Gene transfer of nNOS or eNOS to the penis has been shown to augment erectile responses in aging rats (Champion et al., 1999; Magee et al., 2002). Gene transfer of iNOS to the penis also resulted in enhanced intracavernous pressure (Chancellor et al., 2003). However, despite of these encouraging results, it should be noted that mice with disrupted nNOS or eNOS gene have normal erectile function (Burnett et al., 1996, 2002). Compensatory mechanisms, alternative splicing of the disrupted gene, and/or other unknown mechanisms are possibly involved in the preservation of erectile function in the knockout mice.

3. Adenyl cyclases

Extracellular signals that are transmitted through the cAMP-signaling pathway bind to and activate specific plasma membrane receptors, which, through their coupled G proteins, activate adenyl cyclases (ACs). To date, nine membrane isoforms and one soluble form of mammalian AC have been cloned and characterized (Ishikawa and Homcy, 1997; Hurley, 1998; Smit and Iyengar, 1998; Defer et al., 2000; Hanoune and Defer, 2001; Patel et al., 2001). The structure of a membrane AC includes a short variable amino terminus, followed by six transmembrane spans, a large cytoplasmic domain, a second set of six transmembrane regions, and another large cytoplasmic domain that includes the carboxyl terminus. The overall similarity among the different ACs is roughly 60%: the most conserved sequences are located in the two cytoplasmic domains and range from $50 \pm 90\%$. While different membrane ACs are regulated differently, they are all stimulated by the GTP-bound form of the Gα subunit and are all (except AC9) stimulated by forskolin (Premont et al., 1996; Yan et al., 1998). Moreover, all members of the membrane ACs are inhibited by P-site analogs, which are essentially adenine nucleoside 3' polyphosphates (Desaubry et al., 1996).

Expression of AC2–7 has been demonstrated in VSMCs of various vascular beds (Zhang et al., 1997; Guldemeester et al., 1998; Wong et al., 2001; Ostrom et al., 2002), but the specific functions of individual AC isoforms remain poorly understood. Two reports have looked into the age-related regulation of AC in the vascular system. First, Cohen et al. (1977) examined the activity of particulate (membrane) ACs in broken cell preparations of rat aorta and mesenteric arteries from 3- to 5- and 9- to 13-week-old rats. While the basal AC activity of the mesenteric arteries was four-fold greater than that of the aorta, there was no detectable difference between the two age groups. Stimulation of AC by the GTP analog, 5'-guanylylimidodiphosphate, did not differ with age either. The authors concluded that the decreased vasorelaxation seen in older individuals was not due to reduced AC activity. However, it should be noted that the "old" rats in this study are in fact young adults.

Mader and Alley (1998) studied aortic media membrane preparations from Fischer rats of four age groups ranging from 6 weeks to 24 months. Basal AC activity was found to increase significantly with age. Forskolin-stimulated activity relative to basal tended to be greater in the 6-week and 6-month preparations compared to the 12- and 24-month preparations. G protein activators (GTP, GppNHp, and NaF) produced no age-related decrease in responsiveness. Isoproterenol and prostaglandin E1 did not produce significant age-related changes in responsiveness over basal activity. Because these data do not agree with data obtained with whole vessels, the authors concluded that the aortic media membrane preparations might not be ideal for assessing age-related changes in β-adrenergic responsiveness in vascular smooth muscle.

The cAMP-signaling pathway plays an insignificant role in physiological erection (Martinez-Pineiro et al., 1993; Trigo-Rocha et al., 1993) and few studies assessed AC expression in the erectile tissue. An intact cAMP-signaling pathway that includes demonstrable AC activities has nevertheless been exploited in drug therapy (e.g., prostaglandin E1) for erectile dysfunction (Lue, 2000).

4. Guanyl cyclases

Extracellular signals that are transmitted through the cGMP-signaling pathway bind to and activate GCs, which then catalyze the conversion of GTP to cGMP. In mammals, seven membrane-bound (particulate) GC isoforms (GC-A to GC-G) and one soluble GC (sGC) have been identified (Andreopoulos and Papapetropoulos, 2000; Garbers, 2000; Kobialka and Gorczyca, 2000; Lucas et al., 2000; Potter and Hunter, 2001; Tamura et al., 2001; Wedel and Garbers, 2001; Tremblay et al., 2002). Based on their ligand specificity, the particulate GCs (pGCs) have been classified as (1) natriuretic peptide receptors (GC-A and GC-B) that are activated by natriuretic peptides, including ANP, BNP, and CNP,

(2) intestinal peptide receptor (GC-C) that is activated by intestinal peptides, including guanylin, uroguanylin, and lymphoguanylin, and (3) orphan receptors (GC-D, -E, -F, and -G), the ligands for which have yet to be identified. The natriuretic peptide receptors play important roles in cardiovascular functions and diseases. For ease of discussion, they were described together with their ligands earlier (Section 2.6).

sGC is the receptor for NO, which mediates endothelium-dependent relaxation of blood vessels (Ignarro et al., 1987a; Palmer et al., 1987) and nerve-dependent relaxation of corpus cavernosum (Ignarro et al., 1990). sGC is a heterodimeric protein consisting of α and β subunits, each of which exists in two isoforms ($\alpha1$, $\alpha2$, and $\beta1$, $\beta2$) that are encoded by two separate genes (Andreopoulos and Papapetropoulos, 2000). Expression of sGC $\beta1$ mRNA is downregulated by cAMP and AC activator forskolin (Shimouchi et al., 1993). Downregulation of sGC $\alpha1$ and $\beta1$ mRNA and protein has also been shown in cells treated with cAMP- and cGMP-elevating agents (Ujiie et al., 1994; Papapetropoulos et al., 1995; Filippov et al., 1997). More recently, downregulation of sGC $\alpha1$ mRNA and protein was shown in denuded rat aortic strips treated with sGC activator YC-1 (Kloss et al., 2003). It was further demonstrated that decreased expression of sGC was due to destabilization of sGC $\alpha1$ mRNA brought about by the downregulation of HuR protein that normally binds to and stabilizes sGC $\alpha1$ mRNA (Kloss et al., 2003).

Decreased protein expression (both $\alpha1$ and $\beta1$ subunits) and activity of sGC has been observed in the aorta of both young (6 weeks) and old (17 months) spontaneously hypertensive SHR rats when compared with normotensive WKY rats (Ruetten et al., 1999). The authors suggested that downregulation of sGC is an early event in the pathogenesis of hypertension (Ruetten et al., 1999). In another study, it was found that, in the aorta, whereas the $\beta1$ subunit protein level was similar in young (2 months) rats of SHR and WKY rats, it was downregulated 60 and 80% in old (16 months) WKY and SHR rats, respectively (Kloss et al., 2000). Moreover, the mRNA level of either $\alpha1$ or $\beta1$ subunit was similar in young SHR and WKY rats but $\alpha1$ was 2.5-fold lower and $\beta1$ was 4.3-fold lower in old SHR compared with old WKY. The authors concluded that both aging and hypertension decrease sGC expression, and downregulation of sGC may therefore contribute to arterial dysfunction in senescence and chronic hypertension. Witte et al. (2002) investigated sGC expression and activity in the aorta of Goto–Kakizaki rats with type 2 diabetes mellitus, which is frequently associated with arterial hypertension. The maximum activity of sGC was significantly lower in Goto–Kakizaki rats in all age groups studied (5, 15, and 30 weeks) compared with control nondiabetic rats. The authors concluded that a persistent reduction in sGC activity occurred in Goto–Kakizaki rats shortly after weaning and may contribute to the elevation in blood pressure in this strain of genetically diabetic rats.

Arterial stenosis following balloon angioplasty is associated with undesirable proliferation of VSMCs and a large number of studies have examined various strategies to attenuate VSMC proliferation. Perivascular application of YC-1, an sGC activator, immediately following balloon angioplasty significantly raises cGMP level in the vessel wall with concomitant reduction in neointimal area 2 weeks postoperation (Tulis et al., 2000). Chen et al. (2000a) compared the effects of NO donors and cGMP analogs on the growth of aortic smooth muscle cells (AOSMCs) derived from newborn, adult (3 months of age), and old (2 years) rats. NO donor *S*-nitroso-*N*-acetylpenicillamine failed to block DNA synthesis in AOSMCs from old rats but was effective in AOSMCs from newborn and adult rats. However, cGMP analogs were inhibitory in all three AOSMC types, suggesting that cGMP synthesis might be defective with aging. Western blot analysis revealed that AOSMCs from old rats do not express the β subunit of sGC. To confirm the importance of this observation *in vivo*, the authors balloon-injured the carotid arteries of adult and old rats. Whereas sGC was expressed at the same level in the media of injured vessels and uninjured vessels of both groups, its expression in the intima of old rats was reduced by 70% compared with the intima from adult rats. Furthermore, N^{ω}-nitro-L-arginine, an inhibitor of NO synthesis, enhanced the intimal thickening in injured vessels in adult rats but not in old rats. The authors concluded that the loss of NO responsiveness in aged rats is due to the lack of the β subunit of sGC, and this defect contributes to the enhanced intimal thickening in response to injury in old animals.

sGC plays a pivotal role in erectile function since it provides the link between NO and cGMP that represent the extra- and intracellular signaling molecules, respectively, in physiological erection (Andersson, 2001). In animal studies, sGC activator YC-1 has been shown to cause erectile responses (Mizusawa et al., 2002; Hsieh et al., 2003). Messenger RNAs of the $\alpha 1$, $\alpha 2$, $\beta 1$, and $\beta 2$ subunits of sGC have been detected in human corpus cavernosum (Behrends et al., 2000). Immunohistochemical examination found similar sGC expression in the corpus cavernosum of potent and impotent patients (Klotz et al., 2000). While the membrane-bound GC system is not known to play a role in physiological erection, expression of GC-B in human and rat corpus cavernosum and induction of cavernous smooth muscle relaxation by CNP (ligand for GC-B) have been demonstrated recently (Guidone et al., 2002; Kuthe et al., 2003).

5. Protein kinase A

Protein kinase A (PKA), also called as cAMP-dependent kinase (cAK), is the principal receptor for cAMP, and it mediates the vast majority of the cellular effects of cAMP by phosphorylating a wide variety of downstream targets both in the cytoplasmic and the nuclear compartments (see Section 10) (Meinkoth et al., 1993; Walsh and Van Patten, 1994; Cho-Chung et al., 1995; Francis and Corbin, 1999; Skalhegg and Tasken, 2000; Johnson et al., 2001). PKA is composed of two regulatory (R) and two catalytic (C) subunits that form a tetrameric holoenzyme R_2C_2. Binding of cAMP to the R subunits causes the holoenzyme to dissociate into an $R_2(cAMP)_4$ dimer and two free catalytically active C kinase subunits (Smith et al., 1999). The C subunits are encoded by three different genes, Cα, Cβ, and Cγ, while the regulatory subunits are expressed from four different genes, RIα, RIβ, RIIα, and RIIβ (Doskeland et al., 1993; Taylor et al., 1993). Two forms of the PKA holoenzyme exist: type I formed by RIα and RIβ dimers and type II by RIIα and RIIβ dimers. These isozymes may form from either homo- or heterodimers of the R subunits yielding holoenzyme complexes of PKA with a number of combinatorial configurations including RIα_2C$_2$, RIβ_2C$_2$, RIIα_2C$_2$, RIIβ_2C$_2$, and RIαRIβC$_2$ (Tasken et al., 1993). The presence of multiple C subunit genes further adds to the diversity and complexity of the various holoenzyme complexes that differ in biochemical and functional properties as well as patterns of expression and localization (Pawson and Nash, 2000; Skalhegg and Tasken, 2000). These differences among the isozymes contribute to the broad specificity of PKA in a wide variety of physiological processes in response to cAMP signaling.

That RIα and RIβ subunits have different biochemical properties was examined in an early study, in which PKA holoenzyme formed with the RIβ subunit was found to be preferentially activated by low concentrations of cAMP (Cadd et al., 1990). Recent gene knockout studies have confirmed the nonredundant roles of the RIα and RIβ subunits, as targeted disruption of the RIβ gene in mice resulted in hippocampal dysfunction despite a compensatory increase of the RIα protein and unaltered total PKA activity (Brandon et al., 1995; Hensch et al., 1998). Targeted disruption of the RIIβ gene is also compensated by an increase of the RIα subunit (Cummings et al., 1996), resulting in the formation of a novel holoenzyme that binds cAMP more avidly and is more easily activated than the wild-type holoenzyme. The avidity of the new holoenzyme causes elevations of metabolic rate and body temperature, contributing to a lean phenotype. Targeted disruption of the RIIα gene is also compensated by an increase of the RIα subunit (Burton et al., 1997, 1999), resulting in the formation of a novel holoenzyme that is delocalized from the membrane to the cytosol with no detectable change of functions.

In the absence of cAMP binding, the R subunit is an inhibitor of the C subunit. Overexpression of Cα or Cβ in cell culture is compensated by an increase of RIα at the protein but not the RNA level. The increase of RIα upon overexpression of C subunit is due to protein stabilization resulting from holoenzyme formation, as demonstrated by a 4–5-fold longer half-life of the RIα protein when it is incorporated into holoenzyme than when it exists as a free subunit (Spaulding, 1993; Amieux et al., 1997; Amieux and McKnight, 2002). Such a compensatory mechanism serves to illustrate the importance of maintaining a physiologically appropriate ratio between the R and C subunits, as higher R/C ratio has been shown to associate with lower enzyme activity (Spaulding, 1993). In the aorta of older rats (9–11 months of age), the R/C ratio is 45% higher than in the aorta of younger rats (5–6 weeks of age) (Chin and Hoffman, 1990). This may, in part, explain the loss in β-adrenergic agonists-mediated vasorelaxation in older individuals (Chin and Hoffman, 1990). In rat brain, PKA activity increased from newborn to the mature rats but decreased in older rats (Karege et al., 2001). The decrease of PKA activity is thought to relate to the loss of cognitive capacities in old individuals.

In addition to the cAMP-regulated inhibition of the C subunits by the R subunits, there is also a cAMP-independent regulation of PKA activity by protein kinase inhibitor (PKI) proteins. The PKIs comprise three isoforms, PKIα, β, and γ, that are encoded from three distinct genes (Collins and Uhler, 1997). The PKIα isoform is highly expressed in heart, skeletal muscle, cerebral cortex, and cerebellum, the PKIβ isoform in testis, and the PKIγ isoform in heart, skeletal muscle, and testis. Similar to the R subunits, PKI isoforms inhibit the C subunit through interactions within the substrate-binding site of the C subunit (Johnson et al., 2001). In addition to inhibiting C subunit phosphotransferase activity, the PKI isoforms also serve as chaperones for nuclear export of the C subunit and thereby influence the extent of PKA activity in the nucleus (Johnson et al., 2001).

6. Protein kinase G

Protein kinase G (PKG), also called cGMP-dependent kinase (cGK), is the principal receptor and mediator for cGMP signals (Eigenthaler et al., 1999; Francis and Corbin, 1999; Pfeifer et al., 1999; Ruth, 1999; Carvajal et al., 2000; Hofmann et al., 2000; Schlossmann et al., 2003). In mammals, PKG exists in two major forms, PKG-I and PKG-II, which are encoded by two separate genes (Sandberg et al., 1989; Wernet et al., 1989; Uhler, 1993; Jarchau et al., 1994). PKG-II is absent from the cardiovascular system (Lincoln and Cornwell, 1993; Vaandrager and de Jonge, 1996; Lohmann et al., 1997) and will not be discussed further. PKG-I, which is important for regulating vascular smooth muscle tone, exists as α and β isoforms that arise through alternative splicing of the N-terminal region. The two isoforms are often coexpressed (Keilbach et al., 1992) and whether they perform different physiological functions has not been determined. Biochemical analysis has shown that the amino terminus of PKG-Iα interacts specifically with the myosin-binding subunit of myosin phosphatase (Surks et al., 1999), whereas the amino terminus of PKG-Iβ interacts specifically with inositol 1,4,5-trisphosphate receptor-associated cGMP kinase substrate (IRAG), a PKG-I substrate protein (Ammendola et al., 2001). The isoform-specific domain of the β form, but not the α form, was also shown to possess transcriptional activities when artificially fused to the DNA-binding domain of the yeast transcriptional activator GAL4 (Yuasa et al., 2000). Most recently, reconstituted α isoform in PKG-I-deficient vascular smooth muscle cells was shown to be activated by cGMP while reconstituted β isoform was not (Feil et al., 2002).

The PKG-I polypeptide contains three functional domains: the N-terminal, the regulatory, and the catalytic domains. The N-terminal domain has three functions: dimerization, inhibition, and localization. It contains a leucine zipper that holds two PKG-I molecules together as homodimers, it inhibits the catalytic domain in the basal state, and it interacts with specific cellular proteins that anchor PKG-I to different subcellular locations (see next section). The regulatory domain contains two tandem cGMP-binding sites, and the catalytic domain catalyzes the transfer of the γ phosphate from ATP to a serine/threonine residue of the substrate protein. Binding of cGMP to the regulatory domain induces a conformational change that releases the inhibition of the catalytic domain by the N-terminal domain. Deletion of the N-terminal and regulatory domains results in a constitutively active PKG-I that retains substrate specificity, requires no cGMP for activation (Boerth and Lincoln, 1994), and translocates to the nucleus (Collins and Uhler, 1999; Browning et al., 2001). The translocation of the PKG-I catalytic domain to the nucleus is reminiscent of the well-established phenomenon in which activation of PKA results in the dissociation of the catalytic subunits, which then translocate into the nucleus where the catalytic subunits phosphorylate transcription factor CREB (see Sections 5 and 10). Indeed the PKG-I catalytic domain has been shown to activate CRE-dependent gene transcription (Collins and Uhler, 1999). However, whether the intact PKG-I is capable of nuclear translocation is controversial. On one hand, it has been reported that PKG-I translocates to the nucleus where it phosphorylates CREB, thereby mediating the activation of the *fos* promoter by NO (Gudi et al., 1997, 2000), and that PKG-Iβ physically and functionally interacts with the transcriptional regulator TFII-I (Casteel et al., 2002). On the other hand, three independent studies report no nuclear translocation for PKG-I (Collins and Uhler, 1999, Browning et al., 2001; Feil et al., 2002). The disagreement is perhaps due to differences in the host cells, among which VSMCs were used in one of the studies that reported negative findings (Feil et al., 2002).

While many studies have contributed evidence for a critical role of PKG-I in mediating cGMP-regulated vascular smooth muscle relaxation, the proof was provided in a study using targeted disruption of the PKG-I gene (Pfeifer et al., 1998). Mice lacking PKG-I expression as a result of gene disruption are hypertensive at 4–6 weeks of age. Vascular smooth muscle strips isolated from these mice cannot be relaxed by acetylcholine or NO, although both of which increase cGMP levels. Similarly, cavernous smooth muscle strips from these mice cannot be relaxed by agents that raise cGMP levels and these mice have low ability to reproduce, presumably due to erectile dysfunction (Hedlund et al., 2000). Interestingly, the adenyl cyclase activator forskolin was able to raise cAMP levels and subsequently relax the cavernous smooth muscle strips of the PKG-I knockout mice. These observations further affirm the essential role of the cGMP/PKG-I pathway in physiological erectile function.

By comparing the basal and cGMP-stimulated PKG activities in the aorta of spontaneously hypertensive (SHR) and normotensive rats of 4–12 weeks of age, it has been concluded that the control of PKG activity is abnormal in the early life of hypertensive rats (Coquil et al., 1987). PKG-I was found to be expressed at similar levels in aortic smooth muscle cells (AOSMCs) isolated from young (16 weeks of age) and old (28 months of age) rats (Lin et al., 2001c). However, activation of PKG-I by cGMP was much decreased in AOSMCs of old rats (Lin et al., 2001c). Smooth muscle cells isolated from the corpus cavernosum of young and old rats also expressed similar levels of PKG-I, although those from old rats were less activated by cGMP than those from young rats (Lin et al., 2002).

Both cAMP and cGMP are known to inhibit VSMC growth (Garg and Hassid, 1989; Indolfi et al., 2000; Koyama et al., 2001), and adenovirus-mediated expression of constitutively active PKG-I (see above) reduces neointima formation after balloon injury in rats and reduces coronary in-stent restenosis in pigs (Sinnaeve et al., 2002).

While several studies point to the role of PKG-I *activation* in modulating cellular growth (Hofmann et al., 2000), whether PKG-I *expression* also plays a role is less clear. If indeed phosphorylation of downstream targets by PKG-I is associated with the antiproliferative effects of cGMP and cGMP-elevating agents, then PKG-I expression would seem to be a prerequisite. However, a series of papers from Lincoln and associates have shown that PKG-I downregulation is associated with increased VSMC proliferation. The first of such papers showed that AOSMCs grown at low density or repetitively passaged gradually lost PKG-I expression (Cornwell et al., 1994b). Transfection of PKG-I into AOSMCs that have lost PKG-I expression (i.e., of the synthetic phenotype) resulted in an increased production of contractile phenotype marker proteins (Boerth et al., 1997) and restoration of contractility (Brophy et al., 2002). Downregulated PKG-I expression in the proliferating neointimal smooth muscle cells of coronary artery following balloon injury has also been shown (Anderson et al., 2000; Lincoln et al., 2001). On the other hand, PKG-I knockout mice have histologically normal aorta (Pfeifer et al., 1998), and AOSMCs isolated from these mice are indistinguishable from AOSMCs isolated from wild-type mice in the morphology and general growth characteristics (Feil et al., 2002). Conceivably, the cAMP–PKA signaling pathway, which remains intact in the PKG-I knockout mice (Pfeifer et al., 1998) and plays a more significant role than the cGMP–PKG pathway does in antiproliferation (Koyama et al., 2001), is sufficient to keep cellular growth in check. Lending support to this contention is the demonstration that the antiproliferative effects of NO and cGMP involve the activation of PKA but not PKG-I (Cornwell et al., 1994a).

Cornwell et al. (1994b) reported a dramatically reduced PKG-I expression (barely detectable) in rat AOSMCs at passage 6, and thus suggested that these cells might not be appropriate for the investigation of the role of NO and cGMP in VSMC function. However, in repeated experiments we have found little changes in PKG-I expression in rat AOSMCs at passage 1 through 10 (manuscript in preparation). Furthermore, a random survey of papers published between 2001 and 2003, which have used rat AOSMC cultures, identifies the specified passage numbers between 3 and 22, with 5–10 being the average (Delmolino et al., 2001; Watanabe et al., 2001; El Hadri et al., 2002; Hoffmann et al., 2002; Liu et al., 2002; Merrilees et al., 2002; Stegemann and Nerem, 2003). Although these studies did not examine PKG-I expression, they did verify the α-actin expression in the cultured AOSMCs. As such, it appears that cultured AOSMCs between 5 and 10 passages are well accepted as appropriate experimental subjects and do stably express the contractile phenotype marker α-actin as well as PKG-I.

7. Kinase-anchoring proteins

More than 100 different cellular proteins have been identified as substrates for PKA (see Section 10) (Shabb, 2001). The number of known PKG substrates is smaller but is expanding. Most types of cells, including VSMC, express both PKA and PKG and the majority of PKG substrates are also PKA substrates. So, how does the cell determine which protein to be phosphorylated by which kinase? While a satisfactory answer is still lacking, one emerging theory is that PKA, PKG, and their substrates are compartmentalized within the cell and the cyclic nucleotide signals are processed within these microenvironments. Recent studies have shown that compartmentalization of PKA is carried out by specific cellular proteins called A-kinase anchoring proteins (AKAPs) (Edwards and Scott, 2000; Pawson and Nash, 2000; Skalhegg and Tasken, 2000; Feliciello et al., 2001; Barradeau et al., 2002; Griffioen and Thevelein, 2002; Kapiloff, 2002; Michel and Scott, 2002). There is also evidence that G-kinase (i.e., PKG) anchoring proteins (GKAPs) exist (Vo et al., 1998; Yuasa et al., 2000). These anchoring proteins allow PKA, and perhaps PKG, to be efficiently activated by locally produced cyclic nucleotides (by colocalized adenyl and guanyl cyclases), and at the same time, help PKA and PKG to access specific substrates, thereby effecting localized signaling events.

The AKAP family comprises over 50 members that are structurally diverse but functionally similar, and are defined on the basis of their ability to bind to PKA and coprecipitate with the PKA catalytic activity. All AKAPs contain an amphipathic helix of 14–18 residues that function to bind to the N-termini of the PKA-RII dimer. Each AKAP also contains a unique subcellular targeting domain that restricts its location within the cell. AKAPs tether inactive PKA holoenzymes at defined locations within the cell where they are poised to phosphorylate local substrate in response to cAMP activation. More recent studies also show that AKAPs are capable of coordinating other signaling enzymes into multivalent signaling complexes.

As mentioned in Section 5, two forms of PKA exist: type I with RIα and RIβ dimers and type II with RIIα and RIIβ. Type I PKAs are predominantly cytoplasmic while approximately 75% of type II PKAs are associated with specific cellular structures and organelles. The association of type II PKAs with cellular structures is mediated by AKAPs, and most AKAPs are type II PKA-anchoring proteins that interact with the RII subunit. However, there are a few examples where AKAP interaction is not exclusive to the RII subunit of type II PKAs. For instance, AKAPCE was demonstrated to interact specifically with the RI-like subunit of PKA in *Caenorhabditis elegans* (Angelo and Rubin, 1998), and *in vitro* evidence suggests the existence of a family of dual-function AKAPs that bind to both RI and RII (Huang et al., 1999). In fact, AKAP220 immunoprecipitates from human testis contain both PKA-RI and -RII subunits (Reinton et al., 2000). Besides, in RIIα knockout mice the missing RIIα is functionally replaced by RIα, which localizes PKA to L-type calcium channels in skeletal muscle (Burton et al., 1997).

While the AKAP family is well established, whether there is a GKAP counterpart still awaits verification. Possible GKAPs have been seen as discrete protein bands by *in vitro* binding assays with a PKG-II probe or probes containing the N-terminal regulatory region of PKG-II (Vo et al., 1998). These candidate GKAPs do not bind to PKA and appear to be expressed in a tissue-specific manner in the three rat tissues used in the binding assays (i.e., aorta, brain, and intestine). A definitive GKAP has been cloned with a predicted mass of 42 kd, hence called GKAP42 (Yuasa et al., 2000). It binds to PKG-Iα but not PKG-Iβ, PKG-II, or PKA. Its expression is restricted to spermatocytes and early round spermatids. Finally, it is a substrate for *in vitro* phosphorylation by PKG-Iα and PKG-Iβ, but not PKA. *In vivo*, however, it is phosphorylated by PKG-Iα but not PKG-Iβ.

Few studies have looked into the role of AKAPs in VSMCs. Two such studies showed that binding of AKAP to PKA is required for cAMP stimulation of L-type Ca^{2+} and K_{ATP} channels in smooth muscle cells of rabbit portal vein and rat mesenteric arteries, respectively (Zhong et al., 1999; Hayabuchi et al., 2001). A candidate GKAP of ~ 80 kd has been identified in rat aorta in the above-mentioned binding assays (Vo et al., 1998).

8. Cross-talk

There is evidence that cAMP and cGMP cross-activate PKG and PKA, respectively. Cross-activation of PKG by cAMP has been shown in various vascular tissues, including rat aorta, pig coronary artery, and lamb pulmonary artery (Jiang et al., 1992a; Eckly-Michel et al., 1997; Dhanakoti et al., 2000). There is also substantial overlap between the two cyclic nucleotides in their ability to regulate CO_2-induced cerebral vasodilatation of adult rat pial arteries (Wang et al., 1999). Agents that stimulate cAMP production also cause activation of PKG in VSMCs of coronary and pulmonary arteries (White et al., 2000; Barman et al., 2003). Transfected PKG-Iα or PKG-Iβ in COS-7 fibroblasts was also similarly activated by cAMP and cGMP (Lin et al., 2001c). PKG-I in either AOSMCs or CSMCs was activated at similar levels by cAMP and cGMP, with the exception that PKG-I in CSMCs of old rats was less activated by cGMP than by cAMP (Lin et al., 2001c, 2002). An earlier study showed that PKG was activated by cAMP in rat AOSMCs, leading to a reduction in intracellular Ca^{2+} (Lincoln et al., 1990). Overexpression of nitric oxide synthase, which leads to increased levels of NO and cGMP, was also shown to result in cross-activation of PKA (Lincoln et al., 1995).

Whether cross-activation of PKA and PKG plays physiologically relevant roles has been challenged in a study with PKG-I-deficient mice. In these mice, cGMP- but not cAMP-induced relaxation in aortic rings was impaired (Pfeifer et al., 1998). Furthermore, these mice were hypertensive, indicating PKG-I is the specific mediator of cGMP effects in regulating vascular muscle tone. However, in a more recent study with the same PKG-I-deficient mice, NO was able to relax vascular smooth muscle by increasing sGC activity and cGMP production with subsequent activation of PKA (Sausbier et al., 2000). Thus, the newer data once more support the notion that PKA can be activated by cGMP and this can lead to vasorelaxation.

9. Alternative effectors

While PKA and PKG, respectively, serve as principal effectors for cAMP and cGMP, other cyclic nucleotide effectors have been demonstrated recently. At least two PKA-independent cAMP effector systems are known (Chin et al., 2002), with the first consisting of cyclic nucleotide-gated (CNG) channels that conduct both divalent and monovalent cations. These channels form heterotetrameric complexes consisting of two or three different types of subunits (Biel et al., 1999; Kaupp and Seifert, 2002). Their permeability to Ca^{2+} leads to stimulation of calmodulin and calmodulin-dependent kinases, which in turn modulate cAMP production by regulating the activity of ACs and phosphodiesterases. Their permeability to Na^+ and K^+ could alter the membrane potential in electrically active cells.

CNG channels were originally thought to uniquely operate in the sensory signal transduction of retinal and olfactory cells (Finn et al., 1996). They are now identified in a wide variety of tissues including blood vessels (Biel et al., 1994, 1999; Distler et al., 1994; Ding et al., 1997). So far, six genes encoding CNG channels (CNG1-6) have been identified in mammals (Gerstner et al., 2000; Kaupp and Seifert, 2002). The mRNA of CNG1 channel has been detected in heart, kidney, pineal gland, aorta, and mesenteric arteries (Distler et al., 1994; Ding et al., 1997; Minowa et al., 1997; Yao et al., 1999), and a CNG2-like channel has been cloned from rabbit aorta (Biel et al., 1993). In some cells, such as retinal and olfactory cells, CNG channels are activated by cyclic nucleotides (Finn et al., 1996); in others, such as kidney, they are inhibited by cyclic nucleotides (McCoy et al., 1995). A Ca^{2+}-permeable channel has been reported in VSMCs and inhibition of this channel by cGMP represents another pathway through which nitrovasodilators could lower intracellular Ca^{2+} concentration and thus relax blood vessels (Minowa et al., 1997).

The second PKA-independent cAMP effector system consists of the cAMP-regulated guanine nucleotide exchange factors (cAMP-GEFs) that are also called "exchange proteins activated by cAMP (Epac)". These cAMP-GEFs contain cAMP-binding domains very similar to the cAMP-binding domains in the R subunit of PKA (see Section 5) and are exchange factors of the small GTPases Rap1 and Rap2 (de Rooij et al., 1998, 2000; Kawasaki et al., 1998). In the absence of cAMP, the catalytic activity of Epac1 is inhibited by direct interaction between GEF- and cAMP-binding domains. This inhibition is relieved when cAMP binds to the high-affinity cAMP-binding domain of Epac1, allowing Epac1 to convert Rap1 from the inactive GDP-bound to the active GTP-bound form (de Rooij et al., 2000). A Ras GEF (CNrasGEF) has also recently been shown to be activated by cAMP and cGMP (Pham et al., 2000). Thus, cAMP can activate several small G-proteins in a PKA-independent manner. Since the Ras G-protein signals through the mitogen-activated protein (MAP) kinase signaling pathway that has already been shown to cross-talk with the cAMP-signaling pathway (Bornfeldt and Krebs, 1999), the direct coupling of cAMP to Rap1 activation by cAMP-GEFs adds another dimension to the cAMP-signaling pathway.

In addition to CNG channels and CNrasGEF, at least two other PKG-independent effector systems exist for cGMP. The first involves phosphodiesterases (PDEs) whose role in the termination of cyclic nucleotide signaling will be discussed in more detail in Section 12. PDEs are enzymes that catalyze the conversion of cAMP and cGMP to AMP and GMP, respectively. Some PDEs catalyze the hydrolysis of cAMP or cGMP, while the others catalyze both. PDE3 is one of the dual-specificity PDEs, and increased concentration of cGMP has been shown to result in increase of cAMP levels through competition for the same catalytic sites (Maurice and Haslam, 1990; Francis et al., 2001).

Aquaporin-1 (AQP1) is a member of the diverse major intrinsic protein family of water and solute channels (Matsuzaki et al., 2002). AQP1 is known as an osmotic water channel in kidney, brain, vascular system, and other tissues, and recently has been demonstrated to function as a cation channel gated by cGMP (Anthony et al., 2000). Electrophysiology and binding assays implicate direct cGMP binding in the AQP1 C-terminus (Saparov et al., 2001) and sequence similarities with CNG channels support the idea that the AQP1 C-terminus mediates ion channel activation (Boassa and Yool, 2002). In addition, the AQP1 C-terminus exhibits homology with the cGMP-selectivity subdomain of PDEs, particularly that of PDE5 (Boassa and Yool, 2002).

10. PKA substrates

More than 100 different cellular proteins have been identified as physiological substrates of PKA (Shabb, 2001). Greater than 90% (135 out of 145) of these proteins are phosphorylated at a serine residue, and the remainder at a threonine residue. Slightly more than 50% of all recognized targets are phosphorylated at Arg-Arg-X-Ser, in which Ser is the phosphate acceptor. PKA substrate proteins are involved in cellular processes such as intracellular signaling, transcription regulation, cell death (apoptosis) and survival, ion conductance, homeostasis, motility, and various metabolic processes. Some of these proteins as related to vascular smooth muscle functions are discussed below. Additional PKA substrates that are also PKG substrates are discussed in the next section.

1. Heat shock-related protein (HSP20): Phosphorylation of the 20-kd HSP20 is one of the major phosphorylation events associated with cAMP-induced vasorelaxation in carotid arterial smooth muscle (Beall et al., 1997). Phosphorylation of HSP20 at Ser-16 by PKA has been demonstrated *in vitro* (Beall et al., 1999). Incubation of transiently permeabilized strips of carotid arterial smooth muscle with a synthetic HSP20 peptide in which Ser-16 (phosphoHSP20) is phosphorylated inhibits contractile responses to high extracellular KCl and to serotonin (Beall et al., 1999). Transfection of the wild-type but not the Ser-16 mutant into cultured VSMCs leads to relaxation (Woodrum et al., 2003). Transduction of phosphoHSP20 analogs also leads to dose-dependent relaxation of carotid and coronary artery smooth muscle (Flynn et al., 2003; Woodrum et al., 2003). Interestingly, the vasorelaxation induced by phosphorylated HSP20 is independent of the well-established Ca^{2+}/myosin light-chain pathway (Woodrum et al., 1999; Rembold et al., 2000).

2. Phosphodiesterases (PDEs): PDEs are enzymes that catalyze the hydrolysis of
    cAMP and cGMP; therefore, PDEs play critical roles in the termination of cyclic
    nucleotide signaling (see next section). Among the 11 mammalian PDE families,
    PDE3 and PDE4 are known PKA substrates in VSMC (Ekholm et al., 1997; Liu
    and Maurice, 1999; Shakur et al., 2001). Phosphorylated PDE3 and PDE4
    exhibit higher cAMP-catalytic activities than their unphosphorylated counter-
    parts; therefore, phosphorylation of each of these two PDEs is likely to
    constitute a feedback mechanism that helps blunt excessive cAMP signaling
    (Liu and Maurice, 1999; Conti, 2000).
3. cAMP-responsive element-binding protein (CREB): PKA mediates many of the
    long-term effects of cAMP by phosphorylating the ubiquitous transcription
    factor CREB (cAMP-responsive element-binding protein), which binds as a
    dimer to a conserved cAMP-responsive element (CRE), TGACGTCA, in the
    target gene (Mayr and Montminy, 2001). Phosphorylation of CREB at Ser-133
    promotes transcription by recruitment of the coactivator CREB-binding protein
    (CBP) and results in gene activation in a wide variety of cell types including
    VSMCs. For example, phosphorylation of CREB is associated with angiotensin
    II-induced hypertrophy and thrombin-induced proliferation of VSMCs. As
    proliferation and hypertrophy of VSMCs are characteristics of atherosclerotic
    lesions, cAMP signaling via PKA activation and CREB phosphorylation is thus
    implicated in the pathogenesis of atherosclerosis. A study comparing young and
    senescent fibroblasts has identified aging-related decrease of CREB expression
    and its association with attenuated c-fos gene expression in response to
    cAMP-signaling agents, forskolin and prostaglandin E1 (Chin et al., 1996b).

11. PKG substrates

Far fewer PKG-I substrates have been identified compared with PKA
substrates, and the majority of known PKG substrates are also phosphorylated by
PKA. By screening peptide libraries, the optimal target sequence for PKG-I was
determined to be Arg-Lys-X-(Ser/Thr) with Ser or Thr being the phosphate acceptor
(Tegge et al., 1995). It is generally accepted that PKG-I mediates the vasorelaxation
effect of cGMP by phosphorylating ion channels and proteins that regulate myosin
functions. Some of the proven PKG-I substrates as related to vascular smooth
muscle functions are discussed below.

1. Inositol 1,4,5-trisphosphate (IP3) receptor: IP3 receptor mediates the rise of
    intracellular Ca^{2+} concentration and therefore smooth muscle contraction. It is
    phosphorylated by both PKG and PKA at two sites (Komalavilas and Lincoln,
    1994), resulting in inhibition of Ca release with subsequent smooth muscle
    relaxation (Murthy and Zhou, 2003).
2. IP3 receptor-associated PKG-I substrate (IRAG): IRAG colocalizes with
    IP3 receptor and PKG-I and is preferentially phosphorylated by PKG-Iβ
    (Schlossmann et al., 2000). Coexpression of IRAG and PKG-Iβ in the presence
    of cGMP results in inhibition of calcium release.

3. Phospholamban (PLB): PLB is a 5-kd protein that complexes with the sarcoplasmatic Ca^{2+}-ATPase (SERCA) in heart, smooth muscle, and other tissues (Kadambi and Kranias, 1997). Phosphorylation of PLB at Ser-16 by PKA and PKG results in its dissociation from SERCA and thus allowing Ca^{2+} transport from the cytosol to the sarcoplasmatic reticulum. Experiments with PLB knockout mice indicate that PLB can play a major role in modulating smooth muscle intracellular Ca^{2+} but only a minor role in cyclic nucleotide-mediated relaxation (Lalli et al., 1999).

4. Calcium-activated maxi K^+ channel (BK_{Ca}): Phosphorylation of BK_{Ca} channel by PKG leads to increased opening of this channel and decreased intracellular Ca^{2+} (Alioua et al., 1998). Opening of BK_{Ca} channels hyper-polarizes the membrane and closes a number of channels, including L-type calcium channels, thereby reducing Ca^{2+} influx. At physiological concentrations NO increases the activity of BK_{Ca} channels in VSMCs isolated from wild-type but not PKG-I-deficient mice (Sausbier et al., 2000). Expression of BK_{Ca} channel is drastically reduced in the coronary smooth muscle of aging men and rats, implicating an increased risk of coronary spasm and myocardial ischemia in older people (Marijic et al., 2001).

5. Myosin phosphatase (MP): MP, which dephosphorylates myosin light chain (MLC), leading to reduced smooth muscle tone, is a trimeric protein comprising a large (110–130 kd) regulatory myosin-binding subunit (MBS), a 37-kd catalytic subunit, and a 20-kd protein of unknown function. MBS is phospho-rylated by PKG-I (Nakamura et al., 1999; Surks et al., 1999), and interaction between MBS and PKG-I through a leucine zipper is critical for dephospho-rylation of MLC (Nakamura et al., 1999; Surks et al., 1999).

6. Phosphatase inhibitor-1 (PPI-1): In the presence of cGMP, PPI-I is phospho-rylated by PKG-I at Thr-35. It is suggested that activated PPI-1 inhibits MP, leading to smooth muscle relaxation (Tokui et al., 1996). However, no difference in contractility of aorta or portal vein was noted between wild-type and PPI-1 knockout mice (Carr et al., 2001).

7. Phosphodiesterase 5 (PDE5): PDE5, which is principally responsible for the hydrolysis of cGMP in VSMCs, is phosphorylated at Ser-92 by PKG and PKA. Phosphorylated PDE5 has higher cGMP-catalytic activity than unphospho-rylated PDE5, suggesting a feedback mechanism regulating cGMP signaling (Corbin et al., 2000; Mullershausen et al., 2001).

8. GTPase RhoA: GTPase RhoA activates Rho kinase, which phosphorylates and inactivates MP, leading to smooth muscle contraction. Activity of GTPase RhoA is inhibited by PKG through phosphorylation, thus contributing to SM relaxation (Sauzeau et al., 2000).

9. Heat shock-related protein (HSP20): HSP20 is phosphorylated by both PKA (see previous section) and PKG at Ser-16 (Beall et al., 1997, 1999). The possible role of HSP20 phosphorylation in mediating cGMP-induced vasorelaxation is supported by the observation that cGMP does not increase the phosphorylated level of HSP20 in human umbilical SMCs, which are uniquely refractory to cGMP-induced relaxation (Beall et al., 1997; Woodrum et al., 1999).

12. Phosphodiesterases

In each episode of cyclic nucleotide signaling, the increase of intracellular cAMP or cGMP concentration is typically two- to three-fold over the basal level (Francis et al., 1988, 2001; Jiang et al., 1992a; Gopal et al., 2001). Decline of cyclic nucleotide levels occurs rapidly and often during the continued presence of the signaling hormone (Sette et al., 1994; Wyatt et al., 1998; Francis et al., 2001; Mullershausen et al., 2001). Termination of cyclic nucleotide signals is principally carried out by PDEs that catalyze the hydrolysis of cAMP and cGMP to AMP and GMP, respectively. Feedback mechanisms that increase PDE activities and/or expression by the increased cyclic nucleotide level facilitate the degradation of cyclic nucleotides (Vicini and Conti, 1997; Liu and Maurice, 1999; Corbin et al., 2000; Lin et al., 2001a,b).

The superfamily of mammalian PDEs consists of 11 families (PDE1 to PDE11) that are encoded from 21 distinct genes (Conti, 2000; Essayan, 2001; Francis et al., 2001; Mehats et al., 2002; Lin et al., 2003). Each PDE gene usually encodes more than one isoform through alternative splicing or from alternate gene promoters. PDE1, PDE3, PDE4, PDE7, and PDE8 are multigene families, while PDE2, PDE5, PDE9, PDE10, and PDE11 are unigene families. PDE1, PDE2, PDE3, PDE10, and PDE11 hydrolyze both cAMP and cGMP. PDE4, PDE7, and PDE8 hydrolyze cAMP, while PDE5, PDE6, and PDE9 are specific for cGMP. PDE1–5 have been identified in arterial smooth muscles of aorta, lung, and kidney as well as in cultured VSMC (Miyahara et al., 1995; Jackson et al., 1997; Maclean et al., 1997; Mercapide et al., 1999; Kruuse et al., 2001; Pauvert et al., 2002). PDE6 is specifically expressed in photoreceptor cells, and PDE7–11 are more recently identified PDEs whose expression in vascular smooth muscle has not been reported.

In chronic hypoxia-induced pulmonary hypertensive rats, intracellular cyclic nucleotide levels in pulmonary arteries are lower than control rats and the decreases are associated with increased PDE activities (Maclean et al., 1997). Furthermore, while PDE1–5 have been identified in all pulmonary arteries, different PDE isoforms are differentially upregulated in different arterial branches in the lung of chronic hypoxia-induced pulmonary hypertensive rats (Maclean et al., 1997). PDE3, PDE4, and PDE5 are principally responsible for the hydrolysis of cAMP and cGMP in the smooth muscle of bovine main pulmonary artery (Pauvert et al., 2002). Both cAMP and cGMP are known to inhibit arterial SMC proliferation (Rybalkin and Bornfeldt, 1999; Koyama et al., 2001), and elevated levels of cAMP and cGMP reduce neointimal formation after angioplasty in animal models (Indolfi et al., 2000; Tulis et al., 2000). A recent study showed that PDE1C, which hydrolyzes both cAMP and cGMP, is expressed in proliferating but not quiescent human SMCs (Rybalkin et al., 2002). Inhibition of PDE1C by antisense oligonucleotides or a selective PDE1 inhibitor (8MM-IBMX) results in suppression of SMC proliferation. Thus, it is proposed that PDE1C inhibitors may be of therapeutic value for atherosclerosis and/or restenosis (Rybalkin et al., 2002).

Activation of PDE3 via phosphorylation by Akt kinase (protein kinase B) has been postulated to account for autocoid- and growth factor-induced vasoconstriction (Komalavilas et al., 2001). Various growth factors and autocoids activate Akt, which is a direct downstream target of phosphatidylinositol 3-kinase (PI3-kinase) (Hemmings, 1997; Meier and Hemmings, 1999; Brazil and Hemmings, 2001). LY-294002, a specific inhibitor of PI3-kinase, dose dependently inhibits serotonin- and phorbol ester-induced contraction of vascular smooth muscles. Because PDE3 is a target of Akt (Andersen et al., 1998; Ahmad et al., 2000), it is presumed to be phosphorylated as a result of PI3-kinase inhibition in vascular smooth muscle. Phosphorylation of PDE3B results in higher cAMP-catalytic activity and thus contraction of vascular smooth muscle.

PDE5 has been implicated in the regulation of pulmonary vascular resistance and may thus provide a basis for PDE5 inhibitors as treatments for pulmonary hypertension (Hanson et al., 1998; Zhao et al., 2001). Indeed, PDE5-specific inhibitors such as E4010, E4021, T-1032, and sildenafil have been shown to alleviate pulmonary hypertension (Cohen et al., 1996; Hanasato et al., 1999; Weimann et al., 2000; Zhao et al., 2001; Inoue et al., 2002). PDE5 is principally responsible for the elimination of cGMP signaling in the corpus cavernosum (Francis et al., 2001; Corbin et al., 2002) and its specific inhibitors, sildenafil, tadalafil, and vardenafil are highly effective in treating erectile dysfunction by allowing the accumulation of cGMP in the corpus cavernous smooth muscle (Boolell et al., 1996; Saenz de Tejada et al., 2001; Eardley and Cartledge, 2002).

Expression of other PDEs at the RNA or protein level has also been detected in human corpus cavernosum (Küthe et al., 2001; Uckert et al., 2001), suggesting possible cross-talks among different PDEs. Among these non-PDE5 PDEs, PDE3 also plays a prominent role in penile erection (Kuthe et al., 2002), as demonstrated by the erectogenic effect of a PDE3-specific inhibitor milrinone (Qiu et al., 2000). While direct inhibition of PDE5 is the main mechanism through which sildenafil exerts its erectogenic effect, it has been shown that sildenafil also significantly increases cAMP concentration in isolated human cavernous tissue strips (Stief et al., 2000). This effect is thought to involve PDE3 because cGMP, which is accumulated as a result of PDE5 inhibition by sildenafil, is capable of preventing cAMP degradation by competing for the same catalytic sites on the PDE3 molecules (Maurice and Haslam, 1990; Francis et al., 2001).

Incubation of cultured human CSMCs with forskolin or PGE$_1$ produced significant enhancement of cGMP accumulation (Kim et al., 2000). Because forskolin and PGE$_1$ are direct and indirect activators of adenyl cyclase (AC), respectively, their effect on cGMP accumulation was surprising. By showing that cAMP acted as a competitive inhibitor for PDE5, the authors hypothesized that forskolin- or PGE$_1$-induced cGMP accumulation was due to inhibition of PDE5 by increased cAMP concentration that resulted from AC activation (Kim et al., 2000). While it is clear that cAMP does not interact with PDE5 in an allosteric fashion (Kim et al., 2000), the mechanism through which cAMP inhibits PDE5 remains to be elucidated.

Although differential expression of PDE5 has been shown in the pulmonary vasculature during perinatal development (Hanson et al., 1998), to our knowledge there have been no *bona fide* aging-related studies on PDE expression or activity in vascular or cavernous smooth muscle. However, the relaxant potencies of nonselective PDE inhibitors, theophylline and papaverine, on tracheal smooth muscle have been shown to decrease with increasing age (Preuss and Goldie, 1999). Whether this implies an increased PDE expression or activity in the tracheal smooth muscle and whether it is applicable to vascular smooth muscle remain speculative.

13. Concluding remarks and future directions

Altered control of VSM relaxation and/or growth is implicated in many vascular diseases. In this review we have focused our attention on cAMP- and cGMP-mediated VSM relaxation and growth inhibition in normal and diseased states. We covered all of the major steps of the signaling cascades, beginning with the signaling molecules and their receptors, converging onto the cyclases and cyclic nucleotide-dependent kinases, branching into kinase-anchoring proteins and alternative effectors, and finished with cyclic nucleotide phosphodiesterases. In each step the discussion was started with data concerning the normal mechanisms, progressing into disease- and aging-related areas, and finished with data related to erectile function and dysfunction.

While the importance of cyclic nucleotide signaling in vascular smooth muscle function has been abundantly documented, we have recognized the following difficulties and deficiencies in our present understanding of the topics relating to the themes of this chapter. First, multiple isoforms and variants exist within each protein or peptide family and relatively few of them have been investigated in the vascular smooth muscle or in an aging-related setting. Secondly, the heterogeneity of the various vascular beds (e.g., arteries vs. veins) deems generalization of any given findings inappropriate. Thirdly, only a very limited number of studies have looked into aging-related topics. In fact, there have been no reports of aging-related studies in areas of more recently discovered subjects such as the kinase-anchoring proteins and the alternative effectors.

Although we have only limited understanding of the aging-related changes in cyclic nucleotide signaling, the tremendous success of sildenafil in the treatment of erectile dysfunction has heightened the interest in the basic research as well as the therapy for diseases associated with impaired smooth muscle function. The rapid increase of the aging population in the next decade and beyond will bring a huge number of older men and women requiring treatment for these diseases. The complexity of the cyclic nucleotide signaling molecules and their effectors thus provide an even greater opportunity to develop new molecules with better efficacy and fewer side effects for the treatment of these diseases. An exciting era of applying the basic knowledge of cyclic nucleotide signaling to clinical medicine has just begun.

References

Aboseif, S.R., Breza, J., Bosch, R.J., Benard, F., Stief, C.G., Stackl, W., Lue, T.F., Tanagho, E.A., 1989. Local and systemic effects of chronic intracavernous injection of papaverine, prostaglandin E1, and saline in primates. J. Urol. 142, 403–408.

Ahmad, F., Cong, L.N., Stenson Holst, L., Wang, L.M., Rahn Landstrom, T., Pierce, J.H., Quon, M.J., Degerman, E., Manganiello, V.C., 2000. Cyclic nucleotide phosphodiesterase 3B is a downstream target of protein kinase B and may be involved in regulation of effects of protein kinase B on thymidine incorporation in FDCP2 cells. J. Immunol. 164, 4678–4688.

Alderton, W.K., Cooper, C.E., Knowles, R.G., 2001. Nitric oxide synthases: structure, function and inhibition. Biochem. J. 357, 593–615.

Alioua, A., Tanaka, Y., Wallner, M., Hofmann, F., Ruth, P., Meera, P., Toro, L., 1998. The large conductance, voltage dependent, and calcium-sensitive K + channel, Hslo, is a target of cGMP-dependent protein kinase phosphorylation in vivo. J. Biol. Chem. 273, 32950–32956.

Amieux, P.S., McKnight, G.S., 2002. The essential role of RI alpha in the maintenance of regulated PKA activity. Ann. N.Y. Acad. Sci. 968, 75–95.

Amieux, P.S., Cummings, D.E., Motamed, K., Brandon, E.P., Wailes, L.A., Le, K., Idzerda, R.L., McKnight, G.S., 1997. Compensatory regulation of RIalpha protein levels in protein kinase A mutant mice. J. Biol. Chem. 272, 3993–3998.

Ammendola, A., Geiselhoringer, A., Hofmann, F., Schlossmann, J., 2001. Molecular determinants of the interaction between the inositol 1,4,5-trisphosphate receptor-associated cGMP kinase substrate (IRAG) and cGMP kinase Ibeta. J. Biol. Chem. 276, 24153–24159.

Anderson, R.N., 2002. Deaths: leading causes for 2000. Natl. Vital Stat. Rep. 50, 1–85.

Anderson, F.L., Kralios, A.C., Hershberger, R., Bristow, M.R., 1988. Effect of vasoactive intestinal peptide on myocardial contractility and coronary blood flow in the dog: comparison with isoproterenol and forskolin. J. Cardiovasc. Pharmacol. 12, 365–371.

Andersen, C.B., Roth, R.A., Conti, M., 1998. Protein kinase B/Akt induces resumption of meiosis in Xenopus oocytes. J. Biol. Chem. 273, 18705–18708.

Anderson, P.G., Boerth, N.J., Liu, M., McNamara, D.B., Cornwell, T.L., Lincoln, T.M., 2000. Cyclic GMP-dependent protein kinase expression in coronary arterial smooth muscle in response to balloon catheter injury. Arterioscler. Thromb. Vasc. Biol. 20, 2192–2197.

Andersson, K.E., 2001. Neurophysiology/pharmacology of erection. Int. J. Impot. Res. 13 Suppl. 3, S8–S17.

Andersson, K.E., Wagner, G., 1995. Physiology of penile erection. Physiol. Rev. 75, 191–236.

Andrawis, N., Jones, D.S., Abernethy, D.R., 2000. Aging is associated with endothelial dysfunction in the human forearm vasculature. J. Am. Geriatr. Soc. 48, 193–198.

Andreopoulos, S., Papapetropoulos, A., 2000. Molecular aspects of soluble guanylyl cyclase regulation. Gen. Pharmacol. 34, 147–157.

Angelo, R., Rubin, C.S., 1998. Molecular characterization of an anchor protein (AKAPCE) that binds the RI subunit (RCE) of type I protein kinase A from Caenorhabditis elegans. J. Biol. Chem. 273, 14633–14643.

Angulo, J., Cuevas, P., La Fuente, J.M., Pomerol, J.M., Ruiz-Castane, E., Puigvert, A., Gabancho, S., Fernandez, A., Ney, P., Saenz De Tejada, I., 2002. Regulation of human penile smooth muscle tone by prostanoid receptors. Br. J. Pharmacol. 136, 23–30.

Anthony, T.L., Brooks, H.L., Boassa, D., Leonov, S., Yanochko, G.M., Regan, J.W., Yool, A.J., 2000. Cloned human aquaporin-1 is a cyclic GMP-gated ion channel. Mol. Pharmacol. 57, 576–588.

Arias, E., Smith, B.L., 2003. Deaths: preliminary data for 2001. Natl. Vital Stat. Rep. 51, 1–44.

Arnold, W.P., Mittal, C.K., Katsuki, S., Murad, F., 1977. Nitric oxide activates guanylate cyclase and increases guanosine 3′:5′-cyclic monophosphate levels in various tissue preparations. Proc. Natl. Acad. Sci. USA 74, 3203–3207.

Atlas, S.A., Kleinert, H.D., Camargo, M.J., Januszewicz, A., Sealey, J.E., Laragh, J.H., Schilling, J.W., Lewicki, J.A., Johnson, L.K., Maack, T., 1984. Purification, sequencing and synthesis of natriuretic and vasoactive rat atrial peptide. Nature 309, 717–719.

Barman, S.A., Zhu, S., Han, G., White, R.E., 2003. cAMP activates BKCa channels in pulmonary arterial smooth muscle via cGMP-dependent protein kinase. Am. J. Physiol. Lung Cell. Mol. Physiol 284, L1004–1011.

Barr, C.S., Rhodes, P., Struthers, A.D., 1996. C-type natriuretic peptide. Peptides 17, 1243–1251.

Barradeau, S., Imaizumi-Scherrer, T., Weiss, M.C., Faust, D.M., 2002. Intracellular targeting of the type-I alpha regulatory subunit of cAMP-dependent protein kinase. Trends Cardiovasc. Med. 12, 235–241.

Beall, A.C., Kato, K., Goldenring, J.R., Rasmussen, H., Brophy, C.M., 1997. Cyclic nucleotide-dependent vasorelaxation is associated with the phosphorylation of a small heat shock-related protein. J. Biol. Chem. 272, 11283–11287.

Beall, A., Bagwell, D., Woodrum, D., Stoming, T.A., Kato, K., Suzuki, A., Rasmussen, H., Brophy, C.M., 1999. The small heat shock-related protein, HSP20, is phosphorylated on serine 16 during cyclic nucleotide-dependent relaxation. J. Biol. Chem. 274, 11344–11351.

Beevers, D.G., 1998. Beta-blockers for hypertension: time to call a halt. J. Hum. Hypertens. 12, 807–810.

Begonha, R., Moura, D., Guimaraes, S., 1995. Vascular beta-adrenoceptor-mediated relaxation and the tone of the tissue in canine arteries. J. Pharm. Pharmacol. 47, 510–513.

Behrends, S., Steenpass, A., Porst, H., Scholz, H., 2000. Expression of nitric oxide-sensitive guanylyl cyclase subunits in human corpus cavernosum. Biochem. Pharmacol. 59, 713–717.

Bell, D., McDermott, B.J., 1996. Calcitonin gene-related peptide in the cardiovascular system: characterization of receptor populations and their (patho)physiological significance. Pharmacol. Rev. 48, 253–288.

Biel, M., Altenhofen, W., Hullin, R., Ludwig, J., Freichel, M., Flockerzi, V., Dascal, N., Kaupp, U.B., Hofmann, F., 1993. Primary structure and functional expression of a cyclic nucleotide-gated channel from rabbit aorta. FEBS Lett. 329, 134–138.

Biel, M., Zong, X., Distler, M., Bosse, E., Klugbauer, N., Murakami, M., Flockerzi, V., Hofmann, F., 1994. Another member of the cyclic nucleotide-gated channel family, expressed in testis, kidney, and heart. Proc. Natl. Acad. Sci. USA 91, 3505–3509.

Biel, M., Zong, X., Ludwig, A., Sautter, A., Hofmann, F., 1999. Structure and function of cyclic nucleotide-gated channels. Rev. Physiol. Biochem. Pharmacol. 135, 151–171.

Bivalacqua, T.J., Champion, H.C., Abdel-Mageed, A.B., Kadowitz, P.J., Hellstrom, W.J., 2001. Gene transfer of prepro-calcitonin gene-related peptide restores erectile function in the aged rat. Biol. Reprod. 65, 1371–1377.

Boassa, D., Yool, A.J., 2002. A fascinating tail: cGMP activation of aquaporin-1 ion channels. Trends Pharmacol. Sci. 23, 558–562.

Boerth, N.J., Lincoln, T.M., 1994. Expression of the catalytic domain of cyclic GMP-dependent protein kinase in a baculovirus system. FEBS Lett. 342, 255–260.

Boerth, N.J., Dey, N.B., Cornwell, T.L., Lincoln, T.M., 1997. Cyclic GMP-dependent protein kinase regulates vascular smooth muscle cell phenotype. J. Vasc. Res. 34, 245–259.

Boolell, M., Allen, M.J., Ballard, S.A., Gepi-Attee, S., Muirhead, G.J., Naylor, A.M., Osterloh, I.H., Gingell, C., 1996. Sildenafil: an orally active type 5 cyclic GMP-specific phosphodiesterase inhibitor for the treatment of penile erectile dysfunction. Int. J. Impot. Res. 8, 47–52.

Bornfeldt, K.E., Krebs, E.G., 1999. Crosstalk between protein kinase A and growth factor receptor signaling pathways in arterial smooth muscle. Cell Signal. 11, 465–477.

Bosch, R.J., Benard, F., Aboseif, S.R., 1989. Changes in penile hemodynamics after intra cavernous injection of PGE1 and prostaglandin I2 in pig-tailed monkeys. Int. J. Impot. Res. 1, 211–221.

Brain, S.D., Williams, T.J., Tippins, J.R., Morris, H.R., MacIntyre, I., 1985. Calcitonin gene-related peptide is a potent vasodilator. Nature 313, 54–56.

Brandon, E.P., Zhuo, M., Huang, Y.Y., Qi, M., Gerhold, K.A., Burton, K.A., Kandel, E.R., McKnight, G.S., Idzerda, R.L., 1995. Hippocampal long-term depression and depotentiation are defective in mice carrying a targeted disruption of the gene encoding the RI beta subunit of cAMP-dependent protein kinase. Proc. Natl. Acad. Sci. USA 92, 8851–8855.

Brazil, D.P., Hemmings, B.A., 2001. Ten years of protein kinase B signalling: a hard Akt to follow. Trends Biochem. Sci. 26, 657–664.

Breyer, R.M., Kennedy, C.R., Zhang, Y., Breyer, M.D., 2000. Structure-function analyses of eicosanoid receptors. Physiologic and therapeutic implications. Ann. N.Y. Acad. Sci. 905, 221–231.

Brophy, C.M., Woodrum, D.A., Pollock, J., Dickinson, M., Komalavilas, P., Cornwell, T.L., Lincoln, T.M., 2002. cGMP-dependent protein kinase expression restores contractile function in cultured vascular smooth muscle cells. J. Vasc. Res. 39, 95–103.

Browning, D.D., Mc Shane, M., Marty, C., Ye, R.D., 2001. Functional analysis of type 1alpha cGMP-dependent protein kinase using green fluorescent fusion proteins. J. Biol. Chem. 276, 13039–13048.

Brum, J.M., Bove, A.A., Sufan, Q., Reilly, W., Go, V.L., 1986. Action and localization of vasoactive intestinal peptide in the coronary circulation: evidence for nonadrenergic, noncholinergic coronary regulation. J. Am. Coll. Cardiol. 7, 406–413.

Burnett, A.L., Lowenstein, C.J., Bredt, D.S., Chang, T.S., Snyder, S.H., 1992. Nitric oxide: a physiologic mediator of penile erection. Science 257, 401–403.

Burnett, A.L., Nelson, R.J., Calvin, D.C., Liu, J.X., Demas, G.E., Klein, S.L., Kriegsfeld, L.J., Dawson, V.L., Dawson, T.M., Snyder, S.H., 1996. Nitric oxide-dependent penile erection in mice lacking neuronal nitric oxide synthase. Mol. Med. 2, 288–296.

Burnett, A.L., Chang, A.G., Crone, J.K., Huang, P.L., Sezen, S.E., 2002. Noncholinergic penile erection in mice lacking the gene for endothelial nitric oxide synthase. J. Androl. 23, 92–97.

Burton, K.A., Johnson, B.D., Hausken, Z.E., Westenbroek, R.E., Idzerda, R.L., Scheuer, T., Scott, J.D., Catterall, W.A., McKnight, G.S., 1997. Type II regulatory subunits are not required for the anchoring-dependent modulation of Ca2 + channel activity by cAMP-dependent protein kinase. Proc. Natl. Acad. Sci. USA 94, 11067–11072.

Burton, K.A., Treash-Osio, B., Muller, C.H., Dunphy, E.L., McKnight, G.S., 1999. Deletion of type IIalpha regulatory subunit delocalizes protein kinase A in mouse sperm without affecting motility or fertilization. J. Biol. Chem. 274, 24131–24136.

Busto, R., Prieto, J.C., Bodega, G., Zapatero, J., Carrero, I., 2000. Immunohistochemical localization and distribution of VIP/PACAP receptors in human lung. Peptides 21, 265–269.

Cadd, G.G., Uhler, M.D., McKnight, G.S., 1990. Holoenzymes of cAMP-dependent protein kinase containing the neural form of type I regulatory subunit have an increased sensitivity to cyclic nucleotides. J. Biol. Chem. 265, 19502–19506.

Campbell, G.R., Campbell, J.H., 1990. The phenotypes of smooth muscle expressed in human atheroma. Ann. N.Y. Acad. Sci. 598, 143–158.

Carr, A.N., Sutliff, R.L., Weber, C.S., Allen, P.B., Greengard, P., de Lanerolle, P., Kranias, E.G., Paul, R.J., 2001. Is myosin phosphatase regulated in vivo by inhibitor-1? Evidence from inhibitor-1 knockout mice. J. Physiol. 534, 357–366.

Carrier, S., Nagaraju, P., Morgan, D.M., Baba, K., Nunes, L., Lue, T.F., 1997. Age decreases nitric oxide synthase-containing nerve fibers in the rat penis. J. Urol. 157, 1088–1092.

Cartledge, J.J., Eardley, I., Morrison, J.F., 2001. Nitric oxide-mediated corpus cavernosal smooth muscle relaxation is impaired in ageing and diabetes. BJU Int. 87, 394–401.

Carvajal, J.A., Germain, A.M., Huidobro-Toro, J.P., Weiner, C.P., 2000. Molecular mechanism of cGMP-mediated smooth muscle relaxation. J. Cell Physiol. 184, 409–420.

Casteel, D.E., Zhuang, S., Gudi, T., Tang, J., Vuica, M., Desiderio, S., Pilz, R.B., 2002. cGMP-dependent protein kinase I beta physically and functionally interacts with the transcriptional regulator TFII-I. J. Biol. Chem. 277, 32003–32014.

Cernadas, M.R., Sanchez de Miguel, L., Garcia-Duran, M., Gonzalez-Fernandez, F., Millas, I., Monton, M., Rodrigo, J., Rico, L., Fernandez, P., de Frutos, T., Rodriguez-Feo, J.A., Guerra, J., Caramelo, C., Casado, S., Lopez, F., 1998. Expression of constitutive and inducible nitric oxide synthases in the vascular wall of young and aging rats. Circ. Res. 83, 279–286.

Champion, H.C., Bivalacqua, T.J., Hyman, A.L., Ignarro, L.J., Hellstrom, W.J., Kadowitz, P.J., 1999. Gene transfer of endothelial nitric oxide synthase to the penis augments erectile responses in the aged rat. Proc. Natl. Acad. Sci. USA 96, 11648–11652.

Champion, H.C., Bivalacqua, T.J., Toyoda, K., Heistad, D.D., Hyman, A.L., Kadowitz, P.J., 2000. In vivo gene transfer of prepro-calcitonin gene-related peptide to the lung attenuates chronic hypoxia-induced pulmonary hypertension in the mouse. Circulation 101, 923–930.

Chancellor, M.B., Tirney, S., Mattes, C.E., Tzeng, E., Birder, L.A., Kanai, A.J., De Groat, W.C., Huard, J., Yoshimura, N., 2003. Nitric oxide synthase gene transfer for erectile dysfunction in a rat model. BJU Int. 91, 691–696.

Chao, J., Kato, K., Zhang, J.J., Dobrzynski, E., Wang, C., Agata, J., Chao, L., 2001. Human adrenome-dullin gene delivery protects against cardiovascular remodeling and renal injury. Peptides 22, 1731–1737.

Chauhan, S.D., Nilsson, H., Ahluwalia, A., Hobbs, A.J., 2003. Release of C-type natriuretic peptide accounts for the biological activity of endothelium-derived hyperpolarizing factor. Proc. Natl. Acad. Sci. USA 100, 1426–1431.

Chen, H.H., Burnett, J.C., Jr., 1998. C-type natriuretic peptide: the endothelial component of the natriuretic peptide system. J. Cardiovasc. Pharmacol. 32 Suppl. 3, S22–S28.

Chen, H.H., Burnett, J.C., 2000. Natriuretic peptides in the pathophysiology of congestive heart failure. Curr. Cardiol. Rep. 2, 198–205.

Chen, L., Daum, G., Fischer, J.W., Hawkins, S., Bochaton-Piallat, M.L., Gabbiani, G., Clowes, A.W., 2000a. Loss of expression of the beta subunit of soluble guanylyl cyclase prevents nitric oxide-mediated inhibition of DNA synthesis in smooth muscle cells of old rats. Circ. Res. 86, 520–525.

Chen, Y., Cann, M.J., Litvin, T.N., Iourgenko, V., Sinclair, M.L., Levin, L.R., Buck, J., 2000b. Soluble adenylyl cyclase as an evolutionarily conserved bicarbonate sensor. Science 289, 625–628.

Chen, A.F., Ren, J., Miao, C.Y., 2002. Nitric oxide synthase gene therapy for cardiovascular disease. Jpn. J. Pharmacol. 89, 327–336.

Chin, J.H., Hoffman, B.B., 1990. Age-related deficit in beta receptor stimulation of cAMP binding in blood vessels. Mech. Ageing Dev. 53, 111–125.

Chin, J.H., Hiremath, A.N., Hoffman, B.B., 1996a. cAMP signaling mechanisms with aging in rats. Mech. Ageing Dev. 86, 11–26.

Chin, J.H., Okazaki, M., Frazier, J.S., Hu, Z.W., Hoffman, B.B., 1996b. Impaired cAMP-mediated gene expression and decreased cAMP response element binding protein in senescent cells. Am. J. Physiol. 271, C362–C371.

Chin, K.V., Yang, W.L., Ravatn, R., Kita, T., Reitman, E., Vettori, D., Cvijic, M.E., Shin, M., Iacono, L., 2002. Reinventing the wheel of cyclic AMP: novel mechanisms of cAMP signaling. Ann. N.Y. Acad. Sci. 968, 49–64.

Cho-Chung, Y.S., Pepe, S., Clair, T., Budillon, A., Nesterova, M., 1995. cAMP-dependent protein kinase: role in normal and malignant growth. Crit. Rev. Oncol. Hematol. 21, 33–61.

Chou, T.C., Yen, M.H., Li, C.Y., Ding, Y.A., 1998. Alterations of nitric oxide synthase expression with aging and hypertension in rats. Hypertension 31, 643–648.

Clemens, M.G., Forrester, T., 1981. Appearance of adenosine triphosphate in the coronary sinus effluent from isolated working rat heart in response to hypoxia. J. Physiol. 312, 143–158.

Clifton, V.L., Read, M.A., Leitch, I.M., Giles, W.B., Boura, A.L., Robinson, P.J., Smith, R., 1995. Corticotropin-releasing hormone-induced vasodilatation in the human fetal-placental circulation: involvement of the nitric oxide-cyclic guanosine 3′,5′-monophosphate-mediated pathway. J. Clin. Endocrinol. Metab. 80, 2888–2893.

Cohen, M.L., Blume, A.S., Berkowitz, B.A., 1977. Vascular adenylate cyclase: role of age and guanine nucleotide activation. Blood Vessels 14, 25–42.

Cohen, A.H., Hanson, K., Morris, K., Fouty, B., McMurty, I.F., Clarke, W., Rodman, D.M., 1996. Inhibition of cyclic 3′,5′-guanosine monophosphate-specific phosphodiesterase selectively vasodilates the pulmonary circulation in chronically hypoxic rats. J. Clin. Invest. 97, 172–179.

Collins, S.P., Uhler, M.D., 1997. Characterization of PKIgamma, a novel isoform of the protein kinase inhibitor of cAMP-dependent protein kinase. J. Biol. Chem. 272, 18169–18178.

Collins, S.P., Uhler, M.D., 1999. Cyclic AMP- and cyclic GMP-dependent protein kinases differ in their regulation of cyclic AMP response element-dependent gene transcription. J. Biol. Chem. 274, 8391–8404.

Connat, J.L., Busseuil, D., Gambert, S., Ody, M., Tebaldini, M., Gamboni, S., Faivre, B., Quiquerez, A.L., Millet, M., Michaut, P., Rochette, L., 2001. Modification of the rat aortic wall during ageing; possible relation with decrease of peptidergic innervation. Anat. Embryol. (Berl) 204, 455–468.

Conner, A.C., Hay, D.L., Howitt, S.G., Kilk, K., Langel, U., Wheatley, M., Smith, D.M., Poyner, D.R., 2002. Interaction of calcitonin-gene-related peptide with its receptors. Biochem. Soc. Trans. 30, 451–455.

Conti, M., 2000. Phosphodiesterases and cyclic nucleotide signaling in endocrine cells. Mol. Endocrinol. 14, 1317–1327.

Coquil, J.F., Brunelle, G., Leclerc, L., Cuche, L.J., Guedon, J., 1987. Activity of cyclic GMP-dependent protein kinase in aortae from spontaneously hypertensive rats. J. Hypertens. 5, 347–354.

Corbin, J.D., Turko, I.V., Beasley, A., Francis, S.H., 2000. Phosphorylation of phosphodiesterase-5 by cyclic nucleotide-dependent protein kinase alters its catalytic and allosteric cGMP-binding activities. Eur. J. Biochem. 267, 2760–2767.

Corbin, J.D., Francis, S.H., Webb, D.J., 2002. Phosphodiesterase type 5 as a pharmacologic target in erectile dysfunction. Urology 60, 4–11.

Cornwell, T.L., Arnold, E., Boerth, N.J., Lincoln, T.M., 1994a. Inhibition of smooth muscle cell growth by nitric oxide and activation of cAMP-dependent protein kinase by cGMP. Am. J. Physiol. 267, C1405–C1413.

Cornwell, T.L., Soff, G.A., Traynor, A.E., Lincoln, T.M., 1994b. Regulation of the expression of cyclic GMP-dependent protein kinase by cell density in vascular smooth muscle cells. J. Vasc. Res. 31, 330–337.

Crass, M.F., 3rd, Borst, S.E., Scarpace, P.J., 1992. Beta-adrenergic responsiveness in cultured aorta smooth muscle cells. Effects of subculture and aging. Biochem. Pharmacol. 43, 1811–1815.

Cummings, D.E., Brandon, E.P., Planas, J.V., Motamed, K., Idzerda, R.L., McKnight, G.S., 1996. Genetically lean mice result from targeted disruption of the RII beta subunit of protein kinase A. Nature 382, 622–626.

Dahiya, R., Lin, A., Bakircioglu, M.E., Huang, S.T., Lue, T.F., 1997. mRNA and protein expression of nitric oxide synthase and adrenoceptor alpha 1 in young and old rat penile tissues. Br. J. Urol. 80, 300–306.

Dauphin, F., MacKenzie, E.T., 1995. Cholinergic and vasoactive intestinal polypeptidergic innervation of the cerebral arteries. Pharmacol. Ther. 67, 385–417.

Davies, J.M., Williams, K.I., 1984. Endothelial-dependent relaxant effects of vaso-active intestinal polypeptide and arachidonic acid in rat aortic strips. Prostaglandins 27, 195–202.

Davis, K.L., Martin, E., Turko, I.V., Murad, F., 2001. Novel effects of nitric oxide. Annu. Rev. Pharmacol. Toxicol. 41, 203–236.

de la Torre, J.C., 2002. Alzheimer disease as a vascular disorder: nosological evidence. Stroke 33, 1152–1162.

de Rooij, J., Zwartkruis, F.J., Verheijen, M.H., Cool, R.H., Nijman, S.M., Wittinghofer, A., Bos, J.L., 1998. Epac is a Rap1 guanine-nucleotide-exchange factor directly activated by cyclic AMP. Nature 396, 474–477.

de Rooij, J., Rehmann, H., van Triest, M., Cool, R.H., Wittinghofer, A., Bos, J.L., 2000. Mechanism of regulation of the Epac family of cAMP-dependent RapGEFs. J. Biol. Chem. 275, 20829–20836.

Defer, N., Best-Belpomme, M., Hanoune, J., 2000. Tissue specificity and physiological relevance of various isoforms of adenylyl cyclase. Am. J. Physiol. Renal. Physiol. 279, F400–F416.

Delmolino, L.M., Stearns, N.A., Castellot, J.J., Jr., 2001. COP-1, a member of the CCN family, is a heparin-induced growth arrest specific gene in vascular smooth muscle cells. J. Cell Physiol. 188, 45–55.

Desaubry, L., Shoshani, I., Johnson, R.A., 1996. Inhibition of adenylyl cyclase by a family of newly synthesized adenine nucleoside 3'-polyphosphates. J. Biol. Chem. 271, 14028–14034.

Dhanakoti, S.N., Gao, Y., Nguyen, M.Q., Raj, J.U., 2000. Involvement of cGMP-dependent protein kinase in the relaxation of ovine pulmonary arteries to cGMP and cAMP. J. Appl. Physiol. 88, 1637–1642.

Di Nardo, P., Minieri, M., Carbone, A., Maggiano, N., Micheletti, R., Peruzzi, G., Tallarida, G., 1993. Myocardial expression of atrial natriuretic factor gene in early stages of hamster cardiomyopathy. Mol. Cell. Biochem. 125, 179–192.

Ding, C., Potter, E.D., Qiu, W., Coon, S.L., Levine, M.A., Guggino, S.E., 1997. Cloning and widespread distribution of the rat rod-type cyclic nucleotide-gated cation channel. Am. J. Physiol. 272, C1335–C1344.

Dinsmore, W.W., Alderdice, D.K., 1998. Vasoactive intestinal polypeptide and phentolamine mesylate administered by autoinjector in the treatment of patients with erectile dysfunction resistant to other intracavernosal agents. Br. J. Urol. 81, 437–440.

Distler, M., Biel, M., Flockerzi, V., Hofmann, F., 1994. Expression of cyclic nucleotide-gated cation channels in non-sensory tissues and cells. Neuropharmacology 33, 1275–1282.

Djamilian, M., Stief, C.G., Kuczyk, M., Jonas, U., 1993. Followup results of a combination of calcitonin gene-related peptide and prostaglandin E1 in the treatment of erectile dysfunction. J. Urol. 149, 1296–1298.

Docherty, J.R., 1990. Cardiovascular responses in ageing: a review. Pharmacol. Rev. 42, 103–125.

Dohi, Y., Kojima, M., Sato, K., Luscher, T.F., 1995. Age-related changes in vascular smooth muscle and endothelium. Drugs Aging 7, 278–291.

Doskeland, S.O., Maronde, E., Gjertsen, B.T., 1993. The genetic subtypes of cAMP-dependent protein kinase – functionally different or redundant? Biochim. Biophys. Acta 1178, 249–258.

Duncker, D.J., van Zon, N.S., Pavek, T.J., Herlinger, S.K., Bache, R.J., 1995. Endogenous adenosine mediates coronary vasodilation during exercise after K(ATP)+ channel blockade. J. Clin. Invest. 95, 285–295.

Eardley, I., Cartledge, J., 2002. Tadalafil (Cialis) for men with erectile dysfunction. Int. J. Clin. Pract. 56, 300–304.

Eckly-Michel, A., Martin, V., Lugnier, C., 1997. Involvement of cyclic nucleotide-dependent protein kinases in cyclic AMP-mediated vasorelaxation. Br. J. Pharmacol. 122, 158–164.

Edwards, A.S., Scott, J.D., 2000. A-kinase anchoring proteins: protein kinase A and beyond. Curr. Opin. Cell Biol. 12, 217–221.

Eigenthaler, M., Lohmann, S.M., Walter, U., Pilz, R.B., 1999. Signal transduction by cGMP-dependent protein kinases and their emerging roles in the regulation of cell adhesion and gene expression. Rev. Physiol. Biochem. Pharmacol. 135, 173–209.

Ekholm, D., Belfrage, P., Manganiello, V., Degerman, E., 1997. Protein kinase A-dependent activation of PDE4 (cAMP-specific cyclic nucleotide phosphodiesterase) in cultured bovine vascular smooth muscle cells. Biochim. Biophys. Acta 1356, 64–70.

El Hadri, K., Moldes, M., Mercier, N., Andreani, M., Pairault, J., Feve, B., 2002. Semicarbazide-sensitive amine oxidase in vascular smooth muscle cells: differentiation-dependent expression and role in glucose uptake. Arterioscler. Thromb. Vasc. Biol. 22, 89–94.

El-Salhy, M., Sandstrom, O., 1999. How age changes the content of neuroendocrine peptides in the murine gastrointestinal tract. Gerontology 45, 17–22.

Essayan, D.M., 2001. Cyclic nucleotide phosphodiesterases. J. Allergy Clin. Immunol. 108, 671–680.

Fahrenkrug, J., 1989. VIP and autonomic neurotransmission. Pharmacol. Ther. 41, 515–534.

Feil, R., Gappa, N., Rutz, M., Schlossmann, J., Rose, C.R., Konnerth, A., Brummer, S., Kuhbandner, S., Hofmann, F., 2002. Functional reconstitution of vascular smooth muscle cells with cGMP-dependent protein kinase I isoforms. Circ. Res. 90, 1080–1086.

Feliciello, A., Gottesman, M.E., Avvedimento, E.V., 2001. The biological functions of A-kinase anchor proteins. J. Mol. Biol. 308, 99–114.

Filippi, S., Mancini, M., Amerini, S., Bartolini, M., Natali, A., Mancina, R., Forti, G., Ledda, F., Maggi, M., 2000. Functional adenosine receptors in human corpora cavernosa. Int. J. Androl. 23, 210–217.

Filippov, G., Bloch, D.B., Bloch, K.D., 1997. Nitric oxide decreases stability of mRNAs encoding soluble guanylate cyclase subunits in rat pulmonary artery smooth muscle cells. J. Clin. Invest. 100, 942–948.

Finn, J.T., Grunwald, M.E., Yau, K.W., 1996. Cyclic nucleotide-gated ion channels: an extended family with diverse functions. Annu. Rev. Physiol. 58, 395–426.

Flynn, C.R., Komalavilas, P., Tessier, D., Thresher, J., Niederkofler, E.E., Dreiza, C.M., Nelson, R.W., Panitch, A., Joshi, L., Brophy, C.M., 2003. Transduction of biologically active motifs of the small heat shock-related protein, HSP20, leads to relaxation of vascular smooth muscle. FASEB J, 17, 1358–1360.

Ford, G.A., Hoffman, B.B., Vestal, R.E., Blaschke, T.F., 1992. Age-related changes in adenosine and beta-adrenoceptor responsiveness of vascular smooth muscle in man. Br. J. Clin. Pharmacol. 33, 83–87.

Francis, S.H., Corbin, J.D., 1999. Cyclic nucleotide-dependent protein kinases: intracellular receptors for cAMP and cGMP action. Crit. Rev. Clin. Lab. Sci. 36, 275–328.

Francis, S.H., Noblett, B.D., Todd, B.W., Wells, J.N., Corbin, J.D., 1988. Relaxation of vascular and tracheal smooth muscle by cyclic nucleotide analogs that preferentially activate purified cGMP-dependent protein kinase. Mol. Pharmacol. 34, 506–517.

Francis, S.H., Turko, I.V., Corbin, J.D., 2001. Cyclic nucleotide phosphodiesterases: relating structure and function. Prog. Nucleic Acid Res. Mol. Biol. 65, 1–52.

Fraser, N.J., Wise, A., Brown, J., McLatchie, L.M., Main, M.J., Foord, S.M., 1999. The amino terminus of receptor activity modifying proteins is a critical determinant of glycosylation state and ligand binding of calcitonin receptor-like receptor. Mol. Pharmacol. 55, 1054–1059.

Fredholm, B.B., AP, I.J., Jacobson, K.A., Klotz, K.N., Linden, J., 2001. International Union of Pharmacology. XXV. Nomenclature and classification of adenosine receptors. Pharmacol. Rev. 53, 527–552.

Freis, E.D., 1997. Current status of diuretics, beta blockers, alpha-blockers and alpha-beta-blockers in the treatment of hypertension. Med. Clin. North Am. 81, 1305–1317.

Funk, C.D., 2001. Prostaglandins and leukotrienes: advances in eicosanoid biology. Science 294, 1871–1875.

Furchgott, R.F., Zawadzki, J.V., 1980. The obligatory role of endothelial cells in the relaxation of arterial smooth muscle by acetylcholine. Nature 288, 373–376.

Garbers, D.L., 2000. The guanylyl cyclase receptors. Zygote 8 Suppl. 1, S24–S25.

Garcia-Sainz, J.A., Vazquez-Prado, J., del Carmen Medina, L., 2000. Alpha 1-adrenoceptors: function and phosphorylation. Eur. J. Pharmacol. 389, 1–12.

Garg, U.C., Hassid, A., 1989. Nitric oxide-generating vasodilators and 8-bromo-cyclic guanosine monophosphate inhibit mitogenesis and proliferation of cultured rat vascular smooth muscle cells. J. Clin. Invest. 83, 1774–1777.

Gerhard, M., Roddy, M.A., Creager, S.J., Creager, M.A., 1996. Aging progressively impairs endothelium-dependent vasodilation in forearm resistance vessels of humans. Hypertension 27, 849–853.

Gerstner, A., Zong, X., Hofmann, F., Biel, M., 2000. Molecular cloning and functional characterization of a new modulatory cyclic nucleotide-gated channel subunit from mouse retina. J. Neurosci. 20, 1324–1332.

Gopal, V.K., Francis, S.H., Corbin, J.D., 2001. Allosteric sites of phosphodiesterase-5 (PDE5). A potential role in negative feedback regulation of cGMP signaling in corpus cavernosum. Eur. J. Biochem. 268, 3304–3312.

Grace, G.C., Dusting, G.J., Kemp, B.E., Martin, T.J., 1987. Endothelium and the vasodilator action of rat calcitonin gene-related peptide (CGRP). Br. J. Pharmacol. 91, 729–733.

Gray, D.W., Marshall, I., 1992. Human alpha-calcitonin gene-related peptide stimulates adenylate cyclase and guanylate cyclase and relaxes rat thoracic aorta by releasing nitric oxide. Br. J. Pharmacol. 107, 691–696.

Griffioen, G., Thevelein, J.M., 2002. Molecular mechanisms controlling the localisation of protein kinase A. Curr. Genet. 41, 199–207.

Grossman, E., Messerli, F.H., 2002. Why beta-blockers are not cardioprotective in elderly patients with hypertension. Curr. Cardiol. Rep. 4, 468–473.

Gudi, T., Lohmann, S.M., Pilz, R.B., 1997. Regulation of gene expression by cyclic GMP-dependent protein kinase requires nuclear translocation of the kinase: identification of a nuclear localization signal. Mol. Cell. Biol. 17, 5244–5254.

Gudi, T., Casteel, D.E., Vinson, C., Boss, G.R., Pilz, R.B., 2000. NO activation of fos promoter elements requires nuclear translocation of G-kinase I and CREB phosphorylation but is independent of MAP kinase activation. Oncogene 19, 6324–6333.

Guidone, G., Muller, D., Vogt, K., Mukhopadhyay, A.K., 2002. Characterization of VIP and PACAP receptors in cultured rat penis corpus cavernosum smooth muscle cells and their interaction with guanylate cyclase-B receptors. Regul. Pept. 108, 63–72.

Guimaraes, S., Moura, D., 1999. Advances in vascular neuroeffector mechanisms. Trends Pharmacol. Sci. 20, 90–93.

Guimaraes, S., Moura, D., 2001. Vascular adrenoceptors: an update. Pharmacol. Rev. 53, 319–356.

Guldemeester, H.A., Stenmark, K.R., Brough, G., Tuder, R.M., Stevens, T., 1998. Role of adenylyl cyclase in proliferation of neonatal pulmonary artery smooth muscle cells. Chest 114, 38S–39S.

Haas, C.A., Seftel, A.D., Razmjouei, K., Ganz, M.B., Hampel, N., Ferguson, K., 1998. Erectile dysfunction in aging: upregulation of endothelial nitric oxide synthase. Urology 51, 516–522.

Hadhazy, P., Malomvolgyi, B., Magyar, K., 1988. Endogenous prostanoids and arterial contractility. Prostaglandins. Leukot. Essent. Fatty Acids 32, 175–185.

Hanafy, K.A., Krumenacker, J.S., Murad, F., 2001. NO, nitrotyrosine, and cyclic GMP in signal transduction. Med. Sci. Monit. 7, 801–819.

Hanasato, N., Oka, M., Muramatsu, M., Nishino, M., Adachi, H., Fukuchi, Y., 1999. E-4010, a selective phosphodiesterase 5 inhibitor, attenuates hypoxic pulmonary hypertension in rats. Am. J. Physiol. 277, L225–L232.

Hanoune, J., Defer, N., 2001. Regulation and role of adenylyl cyclase isoforms. Annu. Rev. Pharmacol. Toxicol. 41, 145–174.

Hanson, K.A., Burns, F., Rybalkin, S.D., Miller, J.W., Beavo, J., Clarke, W.R., 1998. Developmental changes in lung cGMP phosphodiesterase-5 activity, protein, and message. Am. J. Respir. Crit. Care Med. 158, 279–288.

Harada, K., Ohashi, K., Kumagai, Y., Ohmori, M., Fujimura, A., 1996. Influence of age on venodilator effect of isoproterenol and amrinone. Eur. J. Clin. Pharmacol. 50, 37–40.

Hasegawa, K., Fujiwara, H., Doyama, K., Miyamae, M., Fujiwara, T., Suga, S., Mukoyama, M., Nakao, K., Imura, H., Sasayama, S., 1993. Ventricular expression of brain natriuretic peptide in hypertrophic cardiomyopathy. Circulation 88, 372–380.

Hayabuchi, Y., Dart, C., Standen, N.B., 2001. Evidence for involvement of A-kinase anchoring protein in activation of rat arterial K(ATP) channels by protein kinase A. J. Physiol. 536, 421–427.

Hayakawa, H., Hirata, Y., Kakoki, M., Suzuki, Y., Nishimatsu, H., Nagata, D., Suzuki, E., Kikuchi, K., Nagano, T., Kangawa, K., Matsuo, H., Sugimoto, T., Omata, M., 1999. Role of nitric oxide-cGMP pathway in adrenomedullin-induced vasodilation in the rat. Hypertension 33, 689–693.

Hedlund, H., Andersson, K.E., 1985a. Comparison of the responses to drugs acting on adrenoreceptors and muscarinic receptors in human isolated corpus cavernosum and cavernous artery. J. Auton. Pharmacol. 5, 81–88.

Hedlund, H., Andersson, K.E., 1985b. Contraction and relaxation induced by some prostanoids in isolated human penile erectile tissue and cavernous artery. J. Urol. 134, 1245–1250.

Hedlund, P., Aszodi, A., Pfeifer, A., Alm, P., Hofmann, F., Ahmad, M., Fassler, R., Andersson, K.E., 2000. Erectile dysfunction in cyclic GMP-dependent kinase I-deficient mice. Proc. Natl. Acad. Sci. USA 97, 2349–2354.

Hemmings, B.A., 1997. Akt signaling: linking membrane events to life and death decisions. Science 275, 628–630.

Henning, R.J., Sawmiller, D.R., 2001. Vasoactive intestinal peptide: cardiovascular effects. Cardiovasc. Res. 49, 27–37.

Hensch, T.K., Gordon, J.A., Brandon, E.P., McKnight, G.S., Idzerda, R.L., Stryker, M.P., 1998. Comparison of plasticity in vivo and in vitro in the developing visual cortex of normal and protein kinase A RIbeta-deficient mice. J. Neurosci. 18, 2108–2117.

Hinschen, A.K., Rose'Meyer, R.B., Headrick, J.P., 2001. Age-related changes in adenosine-mediated relaxation of coronary and aortic smooth muscle. Am. J. Physiol. Heart Circ. Physiol. 280, H2380–H2389.

Hinschen, A.K., Rose'Meyer, R.B., Headrick, J.P., 2003. Adenosine receptor subtypes mediating coronary vasodilation in rat hearts. J. Cardiovasc. Pharmacol. 41, 73–80.

Hofmann, F., Ammendola, A., Schlossmann, J., 2000. Rising behind NO: cGMP-dependent protein kinases. J. Cell Sci. 113(Pt 10), 1671–1676.

Hoffmann, G., Czechowski, M., Schloesser, M., Schobersberger, W., 2002. Procalcitonin amplifies inducible nitric oxide synthase gene expression and nitric oxide production in vascular smooth muscle cells. Crit. Care Med. 30, 2091–2095.

Hsieh, G.C., O'Neill, A.B., Moreland, R.B., Sullivan, J.P., Brioni, J.D., 2003. YC-1 potentiates the nitric oxide/cyclic GMP pathway in corpus cavernosum and facilitates penile erection in rats. Eur. J. Pharmacol. 458, 183–189.

Huang, L.J., Wang, L., Ma, Y., Durick, K., Perkins, G., Deerinck, T.J., Ellisman, M.H., Taylor, S.S., 1999. NH2-Terminal targeting motifs direct dual specificity A-kinase-anchoring protein 1 (D-AKAP1) to either mitochondria or endoplasmic reticulum. J. Cell Biol. 145, 951–959.

Huang, P.L., Huang, Z., Mashimo, H., Bloch, K.D., Moskowitz, M.A., Bevan, J.A., Fishman, M.C., 1995. Hypertension in mice lacking the gene for endothelial nitric oxide synthase. Nature 377, 239–242.

Hurley, J.H., 1998. The adenylyl and guanylyl cyclase superfamily. Curr. Opin Struct. Biol. 8, 770–777.

Ignarro, L.J., Buga, G.M., Wood, K.S., Byrns, R.E., Chaudhuri, G., 1987a. Endothelium-derived relaxing factor produced and released from artery and vein is nitric oxide. Proc. Natl. Acad. Sci. USA 84, 9265–9269.

Ignarro, L.J., Byrns, R.E., Buga, G.M., Wood, K.S., 1987b. Mechanisms of endothelium-dependent vascular smooth muscle relaxation elicited by bradykinin and VIP. Am. J. Physiol. 253, H1074–H1082.

Ignarro, L.J., Bush, P.A., Buga, G.M., Wood, K.S., Fukuto, J.M., Rajfer, J., 1990. Nitric oxide and cyclic GMP formation upon electrical field stimulation cause relaxation of corpus cavernosum smooth muscle. Biochem. Biophys. Res. Commun. 170, 843–850.

Inagami, T., Misono, K.S., Maki, M., Fukumi, H., Takayanagi, R., Grammer, R.T., Tibbetts, C., Pandey, K., Sugiyama, M., Yabe, Y., et al., 1984. Atrial natriuretic factor: purification of active peptides, cloning of cDNA and determination of structures of active peptides and precursors. J. Hypertens. Suppl. 2, S317–S319.

Indolfi, C., Di Lorenzo, E., Rapacciuolo, A., Stingone, A.M., Stabile, E., Leccia, A., Torella, D., Caputo, R., Ciardiello, F., Tortora, G., Chiariello, M., 2000. 8-chloro-cAMP inhibits smooth muscle cell proliferation in vitro and neointima formation induced by balloon injury in vivo. J. Am. Coll. Cardiol. 36, 288–293.

Inoue, H., Yano, K., Noto, T., Takagi, M., Ikeo, T., Kikkawa, K., 2002. Acute and chronic effects of T-1032, a novel selective phosphodiesterase type 5 inhibitor, on monocrotaline-induced pulmonary hypertension in rats. Biol. Pharm. Bull. 25, 1422–1426.

Iranami, H., Hatano, Y., Tsukiyama, Y., Maeda, H., Mizumoto, K., 1996. A beta-adrenoceptor agonist evokes a nitric oxide-cGMP relaxation mechanism modulated by adenylyl cyclase in rat aorta. Halothane does not inhibit this mechanism. Anesthesiology 85, 1129–1138.

Ishikawa, Y., Homcy, C.J., 1997. The adenylyl cyclases as integrators of transmembrane signal transduction. Circ. Res. 80, 297–304.

Ishiyama, Y., Kitamura, K., Ichiki, Y., Nakamura, S., Kida, O., Kangawa, K., Eto, T., 1993. Hemodynamic effects of a novel hypotensive peptide, human adrenomedullin, in rats. Eur. J. Pharmacol. 241, 271–273.

Itoh, H., Lederis, K.P., Rorstad, O.P., 1990. Relaxation of isolated bovine coronary arteries by vasoactive intestinal peptide. Eur. J. Pharmacol. 181, 199–205.

Iwasaki, H., Eguchi, S., Shichiri, M., Marumo, F., Hirata, Y., 1998. Adrenomedullin as a novel growth-promoting factor for cultured vascular smooth muscle cells: role of tyrosine kinase-mediated mitogen-activated protein kinase activation. Endocrinology 139, 3432–3441.

Jackson, E.K., Mi, Z., Carcillo, J.A., Gillespie, D.G., Dubey, R.K., 1997. Phosphodiesterases in the rat renal vasculature. J. Cardiovasc. Pharmacol. 30, 798–801.

Jarchau, T., Häusler, C., Markert, T., Pöhler, D., Vanderkerckhove, J., De Jonge, H.R., Lohmann, S.M., Walter, U., 1994. Cloning, expression, and in situ localization of rat intestinal cGMP-dependent protein kinase II. Proc. Natl. Acad. Sci. USA 91, 9426–9430.

Jeremy, J.Y., Mikhailidis, D.P., Dandona, P., 1988. Excitatory receptor-prostanoid synthesis coupling in smooth muscle: mediation by calcium, protein kinase C and G proteins. Prostaglandins Leukot. Essent. Fatty Acids 34, 215–227.

Jiang, H., Colbran, J.L., Francis, S.H., Corbin, J.D., 1992a. Direct evidence for cross-activation of cGMP-dependent protein kinase by cAMP in pig coronary arteries. J. Biol. Chem. 267, 1015–1019.

Jiang, H.X., Chen, P.C., Sobin, S.S., Giannotta, S.L., 1992b. Age related alterations in the response of the pial arterioles to adenosine in the rat. Mech. Ageing Dev. 65, 257–276.

Johnson, D.A., Akamine, P., Radzio-Andzelm, E., Madhusudan, M., Taylor, S.S., 2001. Dynamics of cAMP-dependent protein kinase. Chem. Rev. 101, 2243–2270.

Jovanovic, A., Jovanovic, S., Tulic, I., Grbovic, L., 1998. Predominant role for nitric oxide in the relaxation induced by vasoactive intestinal polypeptide in human uterine artery. Mol. Hum. Reprod. 4, 71–76.

Juaneda, C., Dumont, Y., Quirion, R., 2000. The molecular pharmacology of CGRP and related peptide receptor subtypes. Trends Pharmacol. Sci. 21, 432–438.

Kadambi, V.J., Kranias, E.G., 1997. Phospholamban: a protein coming of age. Biochem. Biophys. Res. Commun. 239, 1–5.

Kangawa, K., Matsuo, H., 1984. Purification and complete amino acid sequence of alpha-human atrial natriuretic polypeptide (alpha-hANP). Biochem. Biophys. Res. Commun. 118, 131–139.

Kano, H., Kohno, M., Yasunari, K., Yokokawa, K., Horio, T., Ikeda, M., Minami, M., Hanehira, T., Takeda, T., Yoshikawa, J., 1996. Adrenomedullin as a novel antiproliferative factor of vascular smooth muscle cells. J. Hypertens. 14, 209–213.

Kapiloff, M.S., 2002. Contributions of protein kinase A anchoring proteins to compartmentation of cAMP signaling in the heart. Mol. Pharmacol. 62, 193–199.

Karege, F., Lambercy, C., Schwald, M., Steimer, T., Cisse, M., 2001. Differential changes of cAMP-dependent protein kinase activity and 3H-cAMP binding sites in rat hippocampus during maturation and aging. Neurosci. Lett. 315, 89–92.

Kaupp, U.B., Seifert, R., 2002. Cyclic nucleotide-gated ion channels. Physiol. Rev. 82, 769–824.

Kawasaki, H., Takasaki, K., Saito, A., Goto, K., 1988. Calcitonin gene-related peptide acts as a novel vasodilator neurotransmitter in mesenteric resistance vessels of the rat. Nature 335, 164–167.

Kawasaki, H., Springett, G.M., Mochizuki, N., Toki, S., Nakaya, M., Matsuda, M., Housman, D.E., Graybiel, A.M., 1998. A family of cAMP-binding proteins that directly activate Rap1. Science 282, 2275–2279.

Keilbach, A., Ruth, P., Hofmann, F., 1992. Detection of cGMP dependent protein kinase isozymes by specific antibodies. Eur. J. Biochem. 208, 467–473.

Kelly, R.A., Smith, T.W., 1996. Nitric oxide and nitrovasodilators: similarities, differences, and interactions. Am. J. Cardiol. 77, 2C–7C.

Kiely, E.A., Bloom, S.R., Williams, G., 1989. Penile response to intracavernosal vasoactive intestinal polypeptide alone and in combination with other vasoactive agents. Br. J. Urol. 64, 191–194.

Kim, S.Z., Kim, S.H., Park, J.K., Koh, G.Y., Cho, K.W., 1998. Presence and biological activity of C-type natriuretic peptide-dependent guanylate cyclase-coupled receptor in the penile corpus cavernosum. J. Urol. 159, 1741–1746.

Kim, N.N., Huang, Y., Moreland, R.B., Kwak, S.S., Goldstein, I., Traish, A., 2000. Cross-regulation of intracellular cGMP and cAMP in cultured human corpus cavernosum smooth muscle cells. Mol. Cell. Biol. Res. Commun. 4, 10–14.

Klimaschewski, L., 1997. VIP – a "very important peptide" in the sympathetic nervous system? Anat. Embryol. (Berl) 196, 269–277.

Klinger, M., Freissmuth, M., Nanoff, C., 2002. Adenosine receptors: G protein-mediated signalling and the role of accessory proteins. Cell Signal. 14, 99–108.

Kloss, S., Bouloumie, A., Mulsch, A., 2000. Aging and chronic hypertension decrease expression of rat aortic soluble guanylyl cyclase. Hypertension 35, 43–47.

Kloss, S., Furneaux, H., Mulsch, A., 2003. Post-transcriptional regulation of soluble guanylyl cyclase expression in rat aorta. J. Biol. Chem. 278, 2377–2383.

Klotz, T., Bloch, W., Zimmermann, J., Ruth, P., Engelmann, U., Addicks, K., 2000. Soluble guanylate cyclase and cGMP-dependent protein kinase I expression in the human corpus cavernosum. Int. J. Impot. Res. 12, 157–164.

Kobialka, M., Gorczyca, W.A., 2000. Particulate guanylyl cyclases: multiple mechanisms of activation. Acta Biochim. Pol. 47, 517–528.

Komalavilas, P., Lincoln, T.M., 1994. Phosphorylation of the inositol 1,4,5-trisphosphate receptor by cyclic GMP-dependent protein kinase. J. Biol. Chem. 269, 8701–8707.

Komalavilas, P., Mehta, S., Wingard, C.J., Dransfield, D.T., Bhalla, J., Woodrum, J.E., Molinaro, J.R., Brophy, C.M., 2001. PI3-kinase/Akt modulates vascular smooth muscle tone via cAMP signaling pathways. J. Appl. Physiol. 91, 1819–1827.

Koyama, H., Bornfeldt, K.E., Fukumoto, S., Nishizawa, Y., 2001. Molecular pathways of cyclic nucleotide-induced inhibition of arterial smooth muscle cell proliferation. J. Cell Physiol. 186, 1–10.

Krick, S., Platoshyn, O., Sweeney, M., Kim, H., Yuan, J.X., 2001. Activation of K + channels induces apoptosis in vascular smooth muscle cells. Am. J. Physiol. Cell Physiol. 280, C970–C979.

Kruuse, C., Rybalkin, S.D., Khurana, T.S., Jansen-Olesen, I., Olesen, J., Edvinsson, L., 2001. The role of cGMP hydrolysing phosphodiesterases 1 and 5 in cerebral artery dilatation. Eur. J. Pharmacol. 420, 55–65.

Küthe, A., Wiedenroth, A., Mägert, H.J., Uckert, S., Forssmann, W.G., Stief, C.G., Jonas, U., 2001. Expression of different phosphodiesterase genes in human cavernous smooth muscle. J. Urol. 165, 280–283.

Kuthe, A., Montorsi, F., Andersson, K.E., Stief, C.G., 2002. Phosphodiesterase inhibitors for the treatment of erectile dysfunction. Curr. Opin. Investig. Drugs 3, 1489–1495.

Kuthe, A., Reinecke, M., Uckert, S., Becker, A., David, I., Heitland, A., Stief, C.G., Forssmann, W.G., Magert, H.J., 2003. Expression of guanylyl cyclase B in the human corpus cavernosum penis and the possible involvement of its ligand C-type natriuretic polypeptide in the induction of penile erection. J. Urol. 169, 1918–1922.

Lalli, M.J., Shimizu, S., Sutliff, R.L., Kranias, E.G., Paul, R.J., 1999. [Ca2 +]i homeostasis and cyclic nucleotide relaxation in aorta of phospholamban-deficient mice. Am. J. Physiol. 277, H963–H970.

Larsson, L.I., Edvinsson, L., Fahrenkrug, J., Hakanson, R., Owman, C., Schaffalitzky de Muckadell, O., Sundler, F., 1976. Immunohistochemical localization of a vasodilatory polypeptide (VIP) in cerebrovascular nerves. Brain Res. 113, 400–404.

Levin, R.M., Wein, A.J., 1980. Adrenergic alpha receptors outnumber beta receptors in human penile corpus cavernosum. Invest. Urol. 18, 225–226.

Lin, K.F., Chao, J., Chao, L., 1999. Atrial natriuretic peptide gene delivery reduces stroke-induced mortality rate in Dahl salt-sensitive rats. Hypertension 33, 219–224.

Lin, C., Chow, S., Lau, A., Tu, R., Lue, T.F., 2001a. Regulation of human PDE5A2 intronic promoter by cAMP and cGMP: identification of a critical Sp1-binding site. Biochem. Biophys. Res. Commun. 280, 693–699.

Lin, C.S., Chow, S., Lau, A., Tu, R., Lue, T.F., 2001b. Identification and regulation of human PDE5A gene promoter. Biochem. Biophys. Res. Commun. 280, 684–692.

Lin, C.S., Liu, X., Tu, R., Chow, S., Lue, T.F., 2001c. Age-related decrease of protein kinase G activation in vascular smooth muscle cells. Biochem. Biophys. Res. Commun. 287, 244–248.

Lin, C.S., Liu, X., Chow, S., Lue, T.F., 2002. Cyclic AMP and cyclic GMP activate protein kinase G in cavernosal smooth muscle cells: old age is a negative factor. BJU Int. 89, 576–582.

Lin, C.S., Xin, Z.C., Lin, G., Lue, T.F., 2003. Phosphodiesterases as therapeutic targets. Urology 61, 685–691.

Lincoln, T.M., Cornwell, T.L., 1993. Intracellular cyclic GMP receptor proteins. FASEB J. 7, 328–338.

Lincoln, T.M., Cornwell, T.L., Taylor, A.E., 1990. cGMP-dependent protein kinase mediates the reduction of Ca2 + by cAMP in vascular smooth muscle cells. Am. J. Physiol. 258, C399–C407.

Lincoln, T.M., Komalavilas, P., Boerth, N.J., MacMillan-Crow, L.A., Cornwell, T.L., 1995. cGMP signaling through cAMP- and cGMP-dependent protein kinases. Adv. Pharmacol. 34, 305–322.

Lincoln, T.M., Dey, N., Sellak, H., 2001. Invited review: cGMP-dependent protein kinase signaling mechanisms in smooth muscle: from the regulation of tone to gene expression. J. Appl. Physiol. 91, 1421–1430.

Liu, H., Maurice, D.H., 1999. Phosphorylation-mediated activation and translocation of the cyclic AMP-specific phosphodiesterase PDE4D3 by cyclic AMP-dependent protein kinase and mitogen-activated protein kinases. A potential mechanism allowing for the coordinated regulation of PDE4D activity and targeting. J. Biol. Chem. 274, 10557–10565.

Liu, X.M., Chapman, G.B., Peyton, K.J., Schafer, A.I., Durante, W., 2002. Carbon monoxide inhibits apoptosis in vascular smooth muscle cells. Cardiovasc. Res. 55, 396–405.

Lohmann, S.M., Vaandrager, A.B., Smolenski, A., Walter, U., De Jonge, H.R., 1997. Distinct and specific functions of cGMP-dependent protein kinases. Trends Biochem. Sci. 22, 307–312.

Lowe, D.G., Camerato, T.R., Goeddel, D.V., 1990. cDNA sequence of the human atrial natriuretic peptide clearance receptor. Nucleic Acids Res. 18, 3412.

Lucas, K.A., Pitari, G.M., Kazerounian, S., Ruiz-Stewart, I., Park, J., Schulz, S., Chepenik, K.P., Waldman, S.A., 2000. Guanylyl cyclases and signaling by cyclic GMP. Pharmacol. Rev. 52, 375–414.

Lucia, P., Caiola, S., Coppola, A., Maroccia, E., Belli, M., De Martinis, C., Buongiorno, A., 1996. Effect of age and relation to mortality on serial changes of vasoactive intestinal peptide in acute myocardial infarction. Am. J. Cardiol. 77, 644–646.

Lue, T.F., 2000. Erectile dysfunction. N. Engl. J. Med. 342, 1802–1813.

Lyons, D., Roy, S., Patel, M., Benjamin, N., Swift, C.G., 1997. Impaired nitric oxide-mediated vasodilatation and total body nitric oxide production in healthy old age. Clin. Sci. (Lond) 93, 519–525.

Maclean, M.R., Johnston, E.D., McCulloch, K.M., Pooley, L., Houslay, M.D., Sweeney, G., 1997. Phosphodiesterase isoforms in the pulmonary arterial circulation of the rat: changes in pulmonary hypertension. J. Pharmacol. Exp. Ther. 283, 619–624.

Mader, S.L., 1992. Influence of animal age on the beta-adrenergic system in cultured rat aortic and mesenteric artery smooth muscle cells. J. Gerontol. 47, B32–B36.

Mader, S.L., Alley, P.A., 1998. Age-related changes in adenylyl cyclase activity in rat aorta membranes. Mech. Ageing Dev. 101, 111–118.

Magee, T.R., Ferrini, M., Garban, H.J., Vernet, D., Mitani, K., Rajfer, J., Gonzalez-Cadavid, N.F., 2002. Gene therapy of erectile dysfunction in the rat with penile neuronal nitric oxide synthase. Biol. Reprod. 67, 20–28.

Mantelli, L., Amerini, S., Ledda, F., Forti, G., Maggi, M., 1995. The potent relaxant effect of adenosine in rabbit corpora cavernosa is nitric oxide independent and mediated by A2 receptors. J. Androl. 16, 312–317.

Marala, R.B., Mustafa, S.J., 1993. Direct evidence for the coupling of A2-adenosine receptor to stimulatory guanine nucleotide-binding-protein in bovine brain striatum. J. Pharmacol. Exp. Ther. 266, 294–300.

Marijic, J., Li, Q., Song, M., Nishimaru, K., Stefani, E., Toro, L., 2001. Decreased expression of voltage- and Ca(2+)-activated K(+) channels in coronary smooth muscle during aging. Circ. Res. 88, 210–216.

Marshall, I., 1992. Mechanism of vascular relaxation by the calcitonin gene-related peptide. Ann. N.Y. Acad. Sci. 657, 204–215.

Martinez-Pineiro, L., Trigo-Rocha, F., Hsu, G.L., von Heyden, B., Lue, T.F., Tanagho, E.A., 1993. Cyclic guanosine monophosphate mediates penile erection in the rat. Eur. Urol. 24, 492–499.

Matsuo, H., 2001. Discovery of a natriuretic peptide family and their clinical application. Can. J. Physiol. Pharmacol. 79, 736–740.

Matsushita, H., Chang, E., Glassford, A.J., Cooke, J.P., Chiu, C.P., Tsao, P.S., 2001. eNOS activity is reduced in senescent human endothelial cells: Preservation by hTERT immortalization. Circ. Res. 89, 793–798.

Matsuzaki, T., Tajika, Y., Tserentsoodol, N., Suzuki, T., Aoki, T., Hagiwara, H., Takata, K., 2002. Aquaporins: a water channel family. Anat. Sci. Int. 77, 85–93.

Maurice, D.H., Haslam, R.J., 1990. Molecular basis of the synergistic inhibition of platelet function by nitrovasodilators and activators of adenylate cyclase: inhibition of cyclic AMP breakdown by cyclic GMP. Mol. Pharmacol. 37, 671–681.

Maxwell, A.J., 2002. Mechanisms of dysfunction of the nitric oxide pathway in vascular diseases. Nitric Oxide 6, 101–124.

Mayr, B., Montminy, M., 2001. Transcriptional regulation by the phosphorylation-dependent factor CREB. Nat. Rev. Mol. Cell. Biol. 2, 599–609.

McCoy, D.E., Guggino, S.E., Stanton, B.A., 1995. The renal cGMP-gated cation channel: its molecular structure and physiological role. Kidney Int. 48, 1125–1133.

McLatchie, L.M., Fraser, N.J., Main, M.J., Wise, A., Brown, J., Thompson, N., Solari, R., Lee, M.G., Foord, S.M., 1998. RAMPs regulate the transport and ligand specificity of the calcitonin-receptor-like receptor. Nature 393, 333–339.

Mehats, C., Andersen, C.B., Filopanti, M., Jin, S.L., Conti, M., 2002. Cyclic nucleotide phosphodiesterases and their role in endocrine cell signaling. Trends Endocrinol. Metab. 13, 29–35.

Meier, R., Hemmings, B.A., 1999. Regulation of protein kinase B. J. Recept. Signal Transduct. Res. 19, 121–128.

Meinkoth, J.L., Alberts, A.S., Went, W., Fantozzi, D., Taylor, S.S., Hagiwara, M., Montminy, M., Feramisco, J.R., 1993. Signal transduction through the cAMP-dependent protein kinase. Mol. Cell. Biochem. 127–128, 179–186.

Mercapide, J., Santiago, E., Alberdi, E., Martinez-Irujo, J.J., 1999. Contribution of phosphodiesterase isoenzymes and cyclic nucleotide efflux to the regulation of cyclic GMP levels in aortic smooth muscle cells. Biochem. Pharmacol. 58, 1675–1683.

Merrilees, M.J., Lemire, J.M., Fischer, J.W., Kinsella, M.G., Braun, K.R., Clowes, A.W., Wight, T.N., 2002. Retrovirally mediated overexpression of versican v3 by arterial smooth muscle cells induces tropoelastin synthesis and elastic fiber formation in vitro and in neointima after vascular injury. Circ. Res. 90, 481–487.

Michel, T., Feron, O., 1997. Nitric oxide synthases: which, where, how, and why? J. Clin. Invest. 100, 2146–2152.

Michel, J.J., Scott, J.D., 2002. AKAP mediated signal transduction. Annu. Rev. Pharmacol. Toxicol. 42, 235–257.

Minamino, N., Kangawa, K., Matsuo, H., 2000. Adrenomedullin: a new peptidergic regulator of the vascular function. Clin. Hemorheol. Microcirc. 23, 95–102.

Minowa, T., Miwa, S., Kobayashi, S., Enoki, T., Zhang, X.F., Komuro, T., Iwamuro, Y., Masaki, T., 1997. Inhibitory effect of nitrovasodilators and cyclic GMP on ET-1-activated Ca(2+)-permeable nonselective cation channel in rat aortic smooth muscle cells. Br. J. Pharmacol. 120, 1536–1544.

Miyahara, M., Ito, M., Itoh, H., Shiraishi, T., Isaka, N., Konishi, T., Nakano, T., 1995. Isoenzymes of cyclic nucleotide phosphodiesterase in the human aorta: characterization and the effects of E4021. Eur. J. Pharmacol. 284, 25–33.

Mizusawa, H., Hedlund, P., Brioni, J.D., Sullivan, J.P., Andersson, K.E., 2002. Nitric oxide independent activation of guanylate cyclase by YC-1 causes erectile responses in the rat. J. Urol. 167, 2276–2281.

Morishige, K., Shimokawa, H., Yamawaki, T., Miyata, K., Eto, Y., Kandabashi, T., Yogo, K., Higo, T., Egashira, K., Ueno, H., Takeshita, A., 2000. Local adenovirus-mediated transfer of C-type natriuretic peptide suppresses vascular remodeling in porcine coronary arteries in vivo. J. Am. Coll. Cardiol. 35, 1040–1047.

Moritoki, H., Matsugi, T., Takase, H., Ueda, H., Tanioka, A., 1990. Evidence for the involvement of cyclic GMP in adenosine-induced, age-dependent vasodilatation. Br. J. Pharmacol. 100, 569–575.

Moroi, M., Zhang, L., Yasuda, T., Virmani, R., Gold, H.K., Fishman, M.C., Huang, P.L., 1998. Interaction of genetic deficiency of endothelial nitric oxide, gender, and pregnancy in vascular response to injury in mice. J. Clin. Invest. 101, 1225–1232.

Mullershausen, F., Russwurm, M., Thompson, W.J., Liu, L., Koesling, D., Friebe, A., 2001. Rapid nitric oxide-induced desensitization of the cGMP response is caused by increased activity of phosphodiesterase type 5 paralleled by phosphorylation of the enzyme. J. Cell Biol. 155, 271–278.

Murad, F., 1999. Cellular signaling with nitric oxide and cyclic GMP. Braz. J. Med. Biol. Res. 32, 1317–1327.

Murthy, K.S., Zhou, H., 2003. Selective phosphorylation of the IP3R-I in vivo by cGMP-dependent protein kinase in smooth muscle. Am. J. Physiol. Gastrointest. Liver Physiol. 284, G221–G230.

Nakamura, M., Ichikawa, K., Ito, M., Yamamori, B., Okinaka, T., Isaka, N., Yoshida, Y., Fujita, S., Nakano, T., 1999. Effects of the phosphorylation of myosin phosphatase by cyclic GMP-dependent protein kinase. Cell Signal. 11, 671–676.

Nakamura, T., Saito, Y., Ohyama, Y., Masuda, H., Sumino, H., Kuro-o, M., Nabeshima, Y., Nagai, R., Kurabayashi, M., 2002. Production of nitric oxide, but not prostacyclin, is reduced in klotho mice. Jpn. J. Pharmacol. 89, 149–156.

Narumiya, S., FitzGerald, G.A., 2001. Genetic and pharmacological analysis of prostanoid receptor function. J. Clin. Invest. 108, 25–30.

Newby, A.C., Zaltsman, A.B., 2000. Molecular mechanisms in intimal hyperplasia. J. Pathol. 190, 300–309.

Nishimatsu, H., Hirata, Y., Hayakawa, H., Nagata, D., Satonaka, H., Suzuki, E., Horie, S., Takeuchi, T., Ohta, N., Homma, Y., Minowada, S., Nagai, R., Kawabe, K., Kitamura, T., 2001. Effects of intracavernous administration of adrenomedullin on erectile function in rats. Peptides 22, 1913–1918.

Ogawa, Y., Itoh, H., Nakao, K., 1995. Molecular biology and biochemistry of natriuretic peptide family. Clin. Exp. Pharmacol. Physiol. 22, 49–53.

Olah, M.E., Stiles, G.L., 2000. The role of receptor structure in determining adenosine receptor activity. Pharmacol. Ther. 85, 55–75.

Olesen, J., Thomsen, L.L., Iversen, H., 1994. Nitric oxide is a key molecule in migraine and other vascular headaches. Trends Pharmacol. Sci. 15, 149–153.

Ostrom, R.S., Liu, X., Head, B.P., Gregorian, C., Seasholtz, T.M., Insel, P.A., 2002. Localization of adenylyl cyclase isoforms and G protein-coupled receptors in vascular smooth muscle cells: expression in caveolin-rich and noncaveolin domains. Mol. Pharmacol. 62, 983–992.

Palmer, R.M., Ferrige, A.G., Moncada, S., 1987. Nitric oxide release accounts for the biological activity of endothelium-derived relaxing factor. Nature 327, 524–526.

Palmer, L.S., Valcic, M., Melman, A., Giraldi, A., Wagner, G., Christ, G.J., 1994. Characterization of cyclic AMP accumulation in cultured human corpus cavernosum smooth muscle cells. J. Urol. 152, 1308–1314.

Pan, H.Y., Hoffman, B.B., Pershe, R.A., Blaschke, T.F., 1986. Decline in beta adrenergic receptor-mediated vascular relaxation with aging in man. J. Pharmacol. Exp. Ther. 239, 802–807.

Pansari, K., Gupta, A., Thomas, P., 2002. Alzheimer's disease and vascular factors: facts and theories. Int. J. Clin. Pract. 56, 197–203.

Papapetropoulos, A., Marczin, N., Mora, G., Milici, A., Murad, F., Catravas, J.D., 1995. Regulation of vascular smooth muscle soluble guanylate cyclase activity, mRNA, and protein levels by cAMP-elevating agents. Hypertension 26, 696–704.

Patel, T.B., Du, Z., Pierre, S., Cartin, L., Scholich, K., 2001. Molecular biological approaches to unravel adenylyl cyclase signaling and function. Gene 269, 13–25.

Pauvert, O., Salvail, D., Rousseau, E., Lugnier, C., Marthan, R., Savineau, J.P., 2002. Characterisation of cyclic nucleotide phosphodiesterase isoforms in the media layer of the main pulmonary artery. Biochem. Pharmacol. 63, 1763–1772.

Pawson, T., Nash, P., 2000. Protein-protein interactions define specificity in signal transduction. Genes Dev. 14, 1027–1047.

Pfeifer, A., Klatt, P., Massberg, S., Ny, L., Sausbier, M., Hirneiss, C., Wang, G.X., Korth, M., Aszódi, A., Andersson, K.E., Krombach, F., Mayerhofer, A., Ruth, P., Fässler, R., Hofmann, F., 1998. Defective smooth muscle regulation in cGMP kinase I-deficient mice. EMBO J. 17, 3045–3051.

Pfeifer, A., Ruth, P., Dostmann, W., Sausbier, M., Klatt, P., Hofmann, F., 1999. Structure and function of cGMP-dependent protein kinases. Rev. Physiol. Biochem. Pharmacol. 135, 105–149.

Pham, N., Cheglakov, I., Koch, C.A., de Hoog, C.L., Moran, M.F., Rotin, D., 2000. The guanine nucleotide exchange factor CNrasGEF activates ras in response to cAMP and cGMP. Curr. Biol. 10, 555–558.

Pickard, R.S., Powell, P.H., Zar, M.A., 1993. Evidence against vasoactive intestinal polypeptide as the relaxant neurotransmitter in human cavernosal smooth muscle. Br. J. Pharmacol. 108, 497–500.

Pollman, M.J., Yamada, T., Horiuchi, M., Gibbons, G.H., 1996. Vasoactive substances regulate vascular smooth muscle cell apoptosis. Countervailing influences of nitric oxide and angiotensin II. Circ. Res. 79, 748–756.

Potter, L.R., Hunter, T., 2001. Guanylyl cyclase-linked natriuretic peptide receptors: structure and regulation. J. Biol. Chem. 276, 6057–6060.

Prado, M.A., Evans-Bain, B., Oliver, K.R., Dickerson, I.M., 2001. The role of the CGRP-receptor component protein (RCP) in adrenomedullin receptor signal transduction. Peptides 22, 1773–1781.

Premont, R.T., Matsuoka, I., Mattei, M.G., Pouille, Y., Defer, N., Hanoune, J., 1996. Identification and characterization of a widely expressed form of adenylyl cyclase. J. Biol. Chem. 271, 13900–13907.

Preuss, J.M., Goldie, R.G., 1999. Age-related changes in airway responsiveness to phosphodiesterase inhibitors and activators of adenyl cyclase and guanylyl cyclase. Pulm. Pharmacol. Ther. 12, 237–243.

Qian, J.Y., Haruno, A., Asada, Y., Nishida, T., Saito, Y., Matsuda, T., Ueno, H., 2002. Local expression of C-type natriuretic peptide suppresses inflammation, eliminates shear stress-induced thrombosis, and prevents neointima formation through enhanced nitric oxide production in rabbit injured carotid arteries. Circ. Res. 91, 1063–1069.

Qiu, Y., Kraft, P., Lombardi, E., Clancy, J., 2000. Rabbit corpus cavernosum smooth muscle shows a different phosphodiesterase profile than human corpus cavernosum. J. Urol 164, 882–886.

Rajasekaran, M., Kasyan, A., Jain, A., Kim, S.W., Monga, M., 2002. Altered growth factor expression in the aging penis: the Brown-Norway rat model. J. Androl. 23, 393–399.

Reid, J.L., 1999. Drug treatment – antihypertensive drugs – the present role of beta blockers and alpha blockers. Clin. Exp. Hypertens. 21, 815–821.

Reinton, N., Collas, P., Haugen, T.B., Skalhegg, B.S., Hansson, V., Jahnsen, T., Tasken, K., 2000. Localization of a novel human A-kinase-anchoring protein, hAKAP220, during spermatogenesis. Dev. Biol. 223, 194–204.

Rembold, C.M., Foster, D.B., Strauss, J.D., Wingard, C.J., Eyk, J.E., 2000. cGMP-mediated phosphorylation of heat shock protein 20 may cause smooth muscle relaxation without myosin light chain dephosphorylation in swine carotid artery. J. Physiol. 524(Pt 3), 865–878.

Reubi, J.C., 2000. In vitro evaluation of VIP/PACAP receptors in healthy and diseased human tissues. Clinical implications. Ann. N.Y. Acad. Sci. 921, 1–25.

Ross, R., 1997. Cellular and molecular studies of atherogenesis. Atherosclerosis. 131 Suppl., S3–S4.

Rubattu, S., Volpe, M., 2001. The atrial natriuretic peptide: a changing view. J. Hypertens. 19, 1923–1931.

Rudic, R.D., Shesely, E.G., Maeda, N., Smithies, O., Segal, S.S., Sessa, W.C., 1998. Direct evidence for the importance of endothelium-derived nitric oxide in vascular remodeling. J. Clin. Invest. 101, 731–736.

Ruetten, H., Zabel, U., Linz, W., Schmidt, H.H., 1999. Downregulation of soluble guanylyl cyclase in young and aging spontaneously hypertensive rats. Circ. Res. 85, 534–541.

Ruth, P., 1999. Cyclic GMP-dependent protein kinases: understanding in vivo functions by gene targeting. Pharmacol. Ther. 82, 355–372.

Rybalkin, S.D., Bornfeldt, K.E., 1999. Cyclic nucleotide phosphodiesterases and human arterial smooth muscle cell proliferation. Thromb. Haemost. 82, 424–434.

Rybalkin, S.D., Rybalkina, I., Beavo, J.A., Bornfeldt, K.E., 2002. Cyclic nucleotide phosphodiesterase 1C promotes human arterial smooth muscle cell proliferation. Circ. Res. 90, 151–157.

Saenz de Tejada, I., Goldstein, I., Krane, R.J., 1988. Local control of penile erection. Nerves, smooth muscle, and endothelium. Urol. Clin. North Am. 15, 9–15.

Saenz de Tejada, I., Kim, N., Lagan, I., Krane, R.J., Goldstein, I., 1989. Regulation of adrenergic activity in penile corpus cavernosum. J. Urol. 142, 1117–1121.

Saenz de Tejada, I., Angulo, J., Cuevas, P., Fernandez, A., Moncada, I., Allona, A., Lledo, E., Korschen, H.G., Niewohner, U., Haning, H., Pages, E., Bischoff, E., 2001. The phosphodiesterase inhibitory selectivity and the in vitro and in vivo potency of the new PDE5 inhibitor vardenafil. Int. J. Impot. Res. 13, 282–290.

Saito, Y., Nakao, K., Arai, H., Sugawara, A., Morii, N., Yamada, T., Itoh, H., Shiono, S., Mukoyama, M., Obata, K., et al., 1987. Atrial natriuretic polypeptide (ANP) in human ventricle. Increased gene expression of ANP in dilated cardiomyopathy. Biochem. Biophys. Res. Commun. 148, 211–217.

Sandberg, M., Natarajan, V., Ronander, I., Kalderon, D., Walter, U., Lohmann, S.M., Jahnsen, T., 1989. Molecular cloning and predicted full-length amino acid sequence of the type I beta isozyme of cGMP-dependent protein kinase from human placenta. Tissue distribution and developmental changes in rat. FEBS Lett. 255, 321–329.

Saparov, S.M., Kozono, D., Rothe, U., Agre, P., Pohl, P., 2001. Water and ion permeation of aquaporin-1 in planar lipid bilayers. Major differences in structural determinants and stoichiometry. J. Biol. Chem. 276, 31515–31520.

Sato, I., Murota, S., 1995. Paracrine function of endothelium-derived nitric oxide. Life Sci. 56, 1079–1087.

Sausbier, M., Schubert, R., Voigt, V., Hirneiss, C., Pfeifer, A., Korth, M., Kleppisch, T., Ruth, P., Hofmann, F., 2000. Mechanisms of NO/cGMP-dependent vasorelaxation. Circ. Res. 87, 825–830.

Sauzeau, V., Le Jeune, H., Cario-Toumaniantz, C., Smolenski, A., Lohmann, S.M., Bertoglio, J., Chardin, P., Pacaud, P., Loirand, G., 2000. Cyclic GMP-dependent protein kinase signaling pathway inhibits RhoA-induced Ca2 + sensitization of contraction in vascular smooth muscle. J. Biol. Chem. 275, 21722–21729.

Schlossmann, J., Ammendola, A., Ashman, K., Zong, X., Huber, A., Neubauer, G., Wang, G.X., Allescher, H.D., Korth, M., Wilm, M., Hofmann, F., Ruth, P., 2000. Regulation of intracellular calcium by a signalling complex of IRAG, IP3 receptor and cGMP kinase Ibeta. Nature 404, 197–201.

Schlossmann, J., Feil, R., Hofmann, F., 2003. Signaling through NO and cGMP-dependent protein kinases. Ann. Med. 35, 21–27.

Schmidt, D.T., Ruhlmann, E., Waldeck, B., Branscheid, D., Luts, A., Sundler, F., Rabe, K.F., 2001. The effect of the vasoactive intestinal polypeptide agonist Ro 25-1553 on induced tone in isolated human airways and pulmonary artery. Naunyn Schmiedebergs Arch. Pharmacol. 364, 314–320.

Sette, C., Iona, S., Conti, M., 1994. The short-term activation of a rolipram-sensitive, cAMP-specific phosphodiesterase by thyroid-stimulating hormone in thyroid FRTL-5 cells is mediated by a cAMP-dependent phosphorylation. J. Biol. Chem. 269, 9245–9252.

Shabb, J.B., 2001. Physiological substrates of cAMP-dependent protein kinase. Chem. Rev. 101, 2381–2411.

Shakur, Y., Holst, L.S., Landstrom, T.R., Movsesian, M., Degerman, E., Manganiello, V., 2001. Regulation and function of the cyclic nucleotide phosphodiesterase (PDE3) gene family. Prog. Nucleic Acid Res. Mol. Biol. 66, 241–277.

Shen, Z.J., Lu, Y.L., Chen, Z.D., Chen, F., Chen, Z., 2000. Effects of androgen and ageing on gene expression of vasoactive intestinal polypeptide in rat corpus cavernosum. BJU Int. 86, 133–137.

Shepherd, J., 2001. Issues surrounding age: vascular disease in the elderly. Curr. Opin. Lipidol. 12, 601–609.

Shimouchi, A., Janssens, S.P., Bloch, D.B., Zapol, W.M., Bloch, K.D., 1993. cAMP regulates soluble guanylate cyclase beta 1-subunit gene expression in RFL-6 rat fetal lung fibroblasts. Am. J. Physiol. 265, L456–L461.

Silberbach, M., Roberts, C.T., Jr., 2001. Natriuretic peptide signalling: molecular and cellular pathways to growth regulation. Cell Signal. 13, 221–231.

Simonsen, U., Garcia-Sacristan, A., Prieto, D., 2002. Penile arteries and erection. J. Vasc. Res. 39, 283–303.

Sinnaeve, P., Chiche, J.D., Gilljins, H., Van Pelt, N., Wirthlin, D., Van De Werf, F., Collen, D., Bloch, K.D., Janssens, S., 2002. Overexpression of a constitutively active protein kinase G mutant reduces neointima formation and in-stent restenosis. Circulation 105, 2911–2916.

Skalhegg, B.S., Tasken, K., 2000. Specificity in the cAMP/PKA signaling pathway. Differential expression, regulation, and subcellular localization of subunits of PKA. Front. Biosci. 5, D678–D693.

Smit, M.J., Iyengar, R., 1998. Mammalian adenylyl cyclases. Adv. Second Messenger Phosphoprotein Res. 32, 1–21.

Smith, J.D., McLean, S.D., Nakayama, D.K., 1998. Nitric oxide causes apoptosis in pulmonary vascular smooth muscle cells. J. Surg. Res. 79, 121–127.

Smith, C.M., Radzio-Andzelm, E., Madhusudan, M., Akamine, P., Taylor, S.S., 1999. The catalytic subunit of cAMP-dependent protein kinase: prototype for an extended network of communication. Prog. Biophys. Mol. Biol. 71, 313–341.

Spaulding, S.W., 1993. The ways in which hormones change cyclic adenosine 3′,5′-monophosphate-dependent protein kinase subunits, and how such changes affect cell behavior. Endocr. Rev. 14, 632–650.

Stegemann, J.P., Nerem, R.M., 2003. Altered response of vascular smooth muscle cells to exogenous biochemical stimulation in two- and three-dimensional culture. Exp. Cell Res. 283, 146–155.

Stief, C.G., Wetterauer, U., 1988. Erectile responses to intracavernous papaverine and phentolamine: comparison of single and combined delivery. J. Urol. 140, 1415–1416.

Stief, C.G., Wetterauer, U., Schaebsdau, F.H., Jonas, U., 1991. Calcitonin-gene-related peptide: a possible role in human penile erection and its therapeutic application in impotent patients. J. Urol. 146, 1010–1014.

Stief, C.G., Uckert, S., Becker, A.J., Harringer, W., Truss, M.C., Forssmann, W.G., Jonas, U., 2000. Effects of sildenafil on cAMP and cGMP levels in isolated human cavernous and cardiac tissue. Urology 55, 146–150.

Surks, H.K., Mochizuki, N., Kasai, Y., Georgescu, S.P., Tang, K.M., Ito, M., Lincoln, T.M., Mendelsohn, M.E., 1999. Regulation of myosin phosphatase by a specific interaction with cGMP-dependent protein kinase Ialpha. Science 286, 1583–1587.

Suzuki, T., Yamazaki, T., Yazaki, Y., 2001. The role of the natriuretic peptides in the cardiovascular system. Cardiovasc. Res. 51, 489–494.

Tabrizchi, R., Bedi, S., 2001. Pharmacology of adenosine receptors in the vasculature. Pharmacol. Ther. 91, 133–147.

Taddei, S., Virdis, A., Mattei, P., Ghiadoni, L., Gennari, A., Fasolo, C.B., Sudano, I., Salvetti, A., 1995. Aging and endothelial function in normotensive subjects and patients with essential hypertension. Circulation 91, 1981–1987.

Taddei, S., Virdis, A., Ghiadoni, L., Salvetti, G., Bernini, G., Magagna, A., Salvetti, A., 2001. Age-related reduction of NO availability and oxidative stress in humans. Hypertension 38, 274–279.

Takahashi, Y., Ishii, N., Lue, T.F., Tanagho, E.A., 1992. Effects of adenosine on canine penile erection. J. Urol. 148, 1323–1325.

Takemura, G., Fujiwara, H., Mukoyama, M., Saito, Y., Nakao, K., Kawamura, A., Ishida, M., Kida, M., Uegaito, T., Tanaka, M., et al., 1991. Expression and distribution of atrial natriuretic peptide in human hypertrophic ventricle of hypertensive hearts and hearts with hypertrophic cardiomyopathy. Circulation 83, 181–190.

Tamura, N., Chrisman, T.D., Garbers, D.L., 2001. The regulation and physiological roles of the guanylyl cyclase receptors. Endocr. J. 48, 611–634.

Tanner, F.C., Meier, P., Greutert, H., Champion, C., Nabel, E.G., Luscher, T.F., 2000. Nitric oxide modulates expression of cell cycle regulatory proteins: a cytostatic strategy for inhibition of human vascular smooth muscle cell proliferation. Circulation 101, 1982–1989.

Tasken, K., Skalhegg, B.S., Solberg, R., Andersson, K.B., Taylor, S.S., Lea, T., Blomhoff, H.K., Jahnsen, T., Hansson, V., 1993. Novel isozymes of cAMP-dependent protein kinase exist in human cells due to formation of RI alpha-RI beta heterodimeric complexes. J. Biol. Chem. 268, 21276–21283.

Taylor, S.S., Zheng, J., Radzio-Andzelm, E., Knighton, D.R., Ten Eyck, L.F., Sowadski, J.M., Herberg, F.W., Yonemoto, W.M., 1993. cAMP-dependent protein kinase defines a family of enzymes. Phil. Trans. R. Soc. Lond. B. Biol. Sci. 340, 315–324.

Tegge, W., Frank, R., Hofmann, F., Dostmann, W.R., 1995. Determination of cyclic nucleotide-dependent protein kinase substrate specificity by the use of peptide libraries on cellulose paper. Biochemistry 34, 10569–10577.

Tokui, T., Brozovich, F., Ando, S., Ikebe, M., 1996. Enhancement of smooth muscle contraction with protein phosphatase inhibitor 1: activation of inhibitor 1 by cGMP-dependent protein kinase. Biochem. Biophys. Res. Commun. 220, 777–783.

Tremblay, J., Desjardins, R., Hum, D., Gutkowska, J., Hamet, P., 2002. Biochemistry and physiology of the natriuretic peptide receptor guanylyl cyclases. Mol. Cell. Biochem. 230, 31–47.

Triggle, C.R., Ding, H., 2002. Endothelium-derived hyperpolarizing factor: is there a novel chemical mediator? Clin. Exp. Pharmacol. Physiol. 29, 153–160.

Trigo-Rocha, F., Hsu, G.L., Donatucci, C.F., Lue, T.F., 1993. The role of cyclic adenosine monophosphate, cyclic guanosine monophosphate, endothelium and nonadrenergic, noncholinergic neurotransmission in canine penile erection. J. Urol. 149, 872–877.

Trovati, M., Massucco, P., Mattiello, L., Cavalot, F., Mularoni, E., Hahn, A., Anfossi, G., 1995. Insulin increases cyclic nucleotide content in human vascular smooth muscle cells: a mechanism potentially involved in insulin-induced modulation of vascular tone. Diabetologia 38, 936–941.

Truss, M.C., Becker, A.J., Thon, W.F., Kuczyk, M., Djamilian, M.H., Stief, C.G., Jonas, U., 1994. Intracavernous calcitonin gene-related peptide plus prostaglandin E1: possible alternative to penile implants in selected patients. Eur. Urol. 26, 40–45.

Tsujimoto, G., Lee, C.H., Hoffman, B.B., 1986. Age-related decrease in beta adrenergic receptor-mediated vascular smooth muscle relaxation. J. Pharmacol. Exp. Ther. 239, 411–415.

Tulis, D.A., Durante, W., Peyton, K.J., Chapman, G.B., Evans, A.J., Schafer, A.I., 2000. YC-1, a benzyl indazole derivative, stimulates vascular cGMP and inhibits neointima formation. Biochem. Biophys. Res. Commun. 279, 646–652.

Uckert, S., Kuthe, A., Stief, C.G., Jonas, U., 2001. Phosphodiesterase isoenzymes as pharmacological targets in the treatment of male erectile dysfunction. World J. Urol. 19, 14–22.

Ueno, H., Haruno, A., Morisaki, N., Furuya, M., Kangawa, K., Takeshita, A., Saito, Y., 1997. Local expression of C-type natriuretic peptide markedly suppresses neointimal formation in rat injured arteries through an autocrine/paracrine loop. Circulation 96, 2272–2279.

Uhler, M.D., 1993. Cloning and expression of a novel cyclic GMP-dependent protein kinase from mouse brain. J. Biol. Chem. 268, 13586–13591.

Ujiie, K., Hogarth, L., Danziger, R., Drewett, J.G., Yuen, P.S., Pang, I.H., Star, R.A., 1994. Homologous and heterologous desensitization of a guanylyl cyclase-linked nitric oxide receptor in cultured rat medullary interstitial cells. J. Pharmacol. Exp. Ther. 270, 761–767.

Vaandrager, A.B., de Jonge, H.R., 1996. Signalling by cGMP-dependent protein kinases. Mol. Cell. Biochem. 157, 23–30.

van Rossum, D., Hanisch, U.K., Quirion, R., 1997. Neuroanatomical localization, pharmacological characterization and functions of CGRP, related peptides and their receptors. Neurosci. Biobehav. Rev. 21, 649–678.

Vaziri, N.D., Wang, X.Q., Ni, Z.N., Kivlighn, S., Shahinfar, S., 2002. Effects of aging and AT-1 receptor blockade on NO synthase expression and renal function in SHR. Biochim. Biophys. Acta 1592, 153–161.

Vela, J., Gutierrez, A., Vitorica, J., Ruano, D., 2003. Rat hippocampal GABAergic molecular markers are differentially affected by ageing. J. Neurochem. 85, 368–377.

Vicini, E., Conti, M., 1997. Characterization of an intronic promoter of a cyclic adenosine 3′,5′-monophosphate (cAMP)-specific phosphodiesterase gene that confers hormone and cAMP inducibility. Mol. Endocrinol. 11, 839–850.

Vilches, J.J., Ceballos, D., Verdu, E., Navarro, X., 2002. Changes in mouse sudomotor function and sweat gland innervation with ageing. Auton. Neurosci. 95, 80–87.

Vo, N.K., Gettemy, J.M., Coghlan, V.M., 1998. Identification of cGMP-dependent protein kinase anchoring proteins (GKAPs). Biochem. Biophys. Res. Commun. 246, 831–835.

Volicer, L., West, C.D., Chase, A.R., Greene, L., 1983. Beta-adrenergic receptor sensitivity in cultured vascular smooth muscle cells: effect of age and of dietary restriction. Mech. Ageing Dev. 21, 283–293.

Walsh, D.A., Van Patten, S.M., 1994. Multiple pathway signal transduction by the cAMP-dependent protein kinase. FASEB J. 8, 1227–1236.

Wang, Q., Bryowsky, J., Minshall, R.D., Pelligrino, D.A., 1999. Possible obligatory functions of cyclic nucleotides in hypercapnia-induced cerebral vasodilation in adult rats. Am. J. Physiol. 276, H480–H487.

Watanabe, T., Yoshizumi, M., Akishita, M., Eto, M., Toba, K., Hashimoto, M., Nagano, K., Liang, Y.Q., Ohike, Y., Iijima, K., Sudoh, N., Kim, S., Nakaoka, T., Yamashita, N., Ako, J., Ouchi, Y., 2001. Induction of nuclear orphan receptor NGFI-B gene and apoptosis in rat vascular smooth muscle cells treated with pyrrolidinedithiocarbamate. Arterioscler. Thromb. Vasc. Biol. 21, 1738–1744.

Wedel, B., Garbers, D., 2001. The guanylyl cyclase family at Y2K. Annu. Rev. Physiol. 63, 215–233.

Weimann, J., Ullrich, R., Hromi, J., Fujino, Y., Clark, M.W., Bloch, K.D., Zapol, W.M., 2000. Sildenafil is a pulmonary vasodilator in awake lambs with acute pulmonary hypertension. Anesthesiology 92, 1702–1712.

Wernet, W., Flockerzi, V., Hofmann, F., 1989. The cDNA of the two isoforms of bovine cGMP-dependent protein kinase. FEBS Lett. 251, 191–196.

Werstiuk, E.S., Lee, R.M., 2000. Vascular beta-adrenoceptor function in hypertension and in ageing. Can. J. Physiol. Pharmacol. 78, 433–452.

White, R.E., Darkow, D.J., Lang, J.L., 1995. Estrogen relaxes coronary arteries by opening BKCa channels through a cGMP-dependent mechanism. Circ. Res. 77, 936–942.

White, R.E., Kryman, J.P., El-Mowafy, A.M., Han, G., Carrier, G.O., 2000. cAMP-dependent vasodilators cross-activate the cGMP-dependent protein kinase to stimulate BK(Ca) channel activity in coronary artery smooth muscle cells. Circ. Res. 86, 897–905.

WHO, 2002. The World Health Report 2002.

Wimalawansa, S.J., 1992. Age-related changes in tissue contents of immunoreactive calcitonin gene-related peptide. Aging (Milano) 4, 211–217.

Wimalawansa, S.J., 1997. Amylin, calcitonin gene-related peptide, calcitonin, and adrenomedullin; a peptide superfamily. Crit. Rev. Neurobiol. 11, 167–239.

Wise, H., Jones, R.L., 1996. Focus on prostacyclin and its novel mimetics. Trends Pharmacol. Sci. 17, 17–21.

Wise, A., Watson-Koken, M.A., Rees, S., Lee, M., Milligan, G., 1997. Interactions of the alpha2A-adrenoceptor with multiple Gi-family G-proteins: studies with pertussis toxin-resistant G-protein mutants. Biochem. J. 321(Pt 3), 721–728.

Witte, K., Jacke, K., Stahrenberg, R., Arlt, G., Reitenbach, I., Schilling, L., Lemmer, B., 2002. Dysfunction of soluble guanylyl cyclase in aorta and kidney of Goto-Kakizaki rats: influence of age and diabetic state. Nitric Oxide 6, 85–95.

Wong, S.T., Baker, L.P., Trinh, K., Hetman, M., Suzuki, L.A., Storm, D.R., Bornfeldt, K.E., 2001. Adenylyl cyclase 3 mediates prostaglandin E(2)-induced growth inhibition in arterial smooth muscle cells. J. Biol. Chem. 276, 34206–34212.

Woodman, C.R., Price, E.M., Laughlin, M.H., 2002. Aging induces muscle-specific impairment of endothelium-dependent dilation in skeletal muscle feed arteries. J. Appl. Physiol. 93, 1685–1690.

Woodrum, D.A., Brophy, C.M., Wingard, C.J., Beall, A., Rasmussen, H., 1999. Phosphorylation events associated with cyclic nucleotide-dependent inhibition of smooth muscle contraction. Am. J. Physiol. 277, H931–H939.

Woodrum, D., Pipkin, W., Tessier, D., Komalavilas, P., Brophy, C.M., 2003. Phosphorylation of the heat shock-related protein, HSP20, mediates cyclic nucleotide-dependent relaxation. J. Vasc. Surg. 37, 874–881.

Wyatt, T.A., Naftilan, A.J., Francis, S.H., Corbin, J.D., 1998. ANF elicits phosphorylation of the cGMP phosphodiesterase in vascular smooth muscle cells. Am. J. Physiol. 274, H448–H455.

Yamahara, K., Itoh, H., Chun, T.H., Ogawa, Y., Yamashita, J., Sawada, N., Fukunaga, Y., Sone, M., Yurugi-Kobayashi, T., Miyashita, K., Tsujimoto, H., Kook, H., Feil, R., Garbers, D.L., Hofmann, F., Nakao, K., 2003. Significance and therapeutic potential of the natriuretic peptides/cGMP/cGMP-dependent protein kinase pathway in vascular regeneration. Proc. Natl. Acad. Sci. USA 100, 3404–3409.

Yan, S.Z., Huang, Z.H., Andrews, R.K., Tang, W.J., 1998. Conversion of forskolin-insensitive to forskolin-sensitive (mouse-type IX) adenylyl cyclase. Mol. Pharmacol. 53, 182–187.

Yao, X., Leung, P.S., Kwan, H.Y., Wong, T.P., Fong, M.W., 1999. Rod-type cyclic nucleotide-gated cation channel is expressed in vascular endothelium and vascular smooth muscle cells. Cardiovasc. Res. 41, 282–290.

Yasue, H., Obata, K., Okumura, K., Kurose, M., Ogawa, H., Matsuyama, K., Jougasaki, M., Saito, Y., Nakao, K., Imura, H., 1989. Increased secretion of atrial natriuretic polypeptide from the left ventricle in patients with dilated cardiomyopathy. J. Clin. Invest. 83, 46–51.

Yuasa, K., Omori, K., Yanaka, N., 2000. Binding and phosphorylation of a novel male germ cell-specific cGMP-dependent protein kinase-anchoring protein by cGMP-dependent protein kinase Ialpha. J. Biol. Chem. 275, 4897–4905.

Zeiher, A.M., Drexler, H., Saurbier, B., Just, H., 1993. Endothelium-mediated coronary blood flow modulation in humans. Effects of age, atherosclerosis, hypercholesterolemia, and hypertension. J. Clin. Invest. 92, 652–662.

Zhang, J., Sato, M., Duzic, E., Kubalak, S.W., Lanier, S.M., Webb, J.G., 1997. Adenylyl cyclase isoforms and vasopressin enhancement of agonist-stimulated cAMP in vascular smooth muscle cells. Am. J. Physiol. 273, H971–H980.

Zhao, L., Mason, N.A., Morrell, N.W., Kojonazarov, B., Sadykov, A., Maripov, A., Mirrakhimov, M.M., Aldashev, A., Wilkins, M.R., 2001. Sildenafil inhibits hypoxia-induced pulmonary hypertension. Circulation 104, 424–428.

Zhong, H., Minneman, K.P., 1999. Alpha1-adrenoceptor subtypes. Eur. J. Pharmacol. 375, 261–276.

Zhong, J., Hume, J.R., Keef, K.D., 1999. Anchoring protein is required for cAMP-dependent stimulation of L-type Ca(2+) channels in rabbit portal vein. Am. J. Physiol. 277, C840–C844.

Zhou, J.N., Swaab, D.F., 1999. Activation and degeneration during aging: a morphometric study of the human hypothalamus. Microsc. Res. Tech. 44, 36–48.

Zhu, B.S., Blessing, W.W., Gibbins, I.L., 1997. Parasympathetic innervation of cephalic arteries in rabbits: comparison with sympathetic and sensory innervation. J. Comp. Neurol. 389, 484–495.

Zippin, J.H., Levin, L.R., Buck, J., 2001. CO(2)/HCO(3)(−)-responsive soluble adenylyl cyclase as a putative metabolic sensor. Trends Endocrinol. Metab. 12, 366–370.

Insulin signaling in the aging nervous system

Eduardo M. Rocha[a], Maria Luiza A. Fernandes[b] and
Lício A. Velloso[a]

[a]*Department of Internal Medicine, State University of Campinas, Brazil.*
[b]*Department of Pathology, State University of Rio de Janeiro, Brazil.*
Correspondence address: L. A. Velloso, DCM-FCM UNICAMP,
13083-970 Campinas, SP, Brazil. Fax: +55-19-3788-8950.
E-mail address: lavelloso@fcm.unicamp.br

Contents

1. Introduction

Insulin action plays a central role in the control of several metabolic and growth-related responses by most, if not all, mammalian cells (Pessin and Saltiel, 2000). The tight regulation of blood glucose levels exerted by insulin is perhaps the most remarkable and well-characterized effect of this pancreatic hormone (Saltiel and Kahn, 2001). Despite variations in fuel availability that might occur following a meal or after a long period of fasting, the level of glucose in plasma remains in a narrow range between 4.0 and 7.0 mM. This control is achieved by the rate of glucose absorption in the gut, uptake, and metabolism by all cells in the body, and production by the liver (Saltiel and Kahn, 2001). Insulin controls glucose homeostasis by inducing its uptake by muscle and adipose tissue, and by inhibiting its output by the liver. In muscle and fat, insulin promotes glucose transporter GLUT4 migration from an intracellular pool to the cell surface (Klip and Paquet, 1990) and, through this mechanism, regulates glucose entrance into the cells. In neural tissues, however, glucose uptake is independent of insulin action

Advances in Cell Aging and Gerontology, vol. 16, 107–132
DOI: 10.1016/S1566-3124(04)16005-7

(Bell et al., 1990). This may be one of the reasons why researchers have taken a long time to focus their attention upon the roles of insulin in the brain.

Insulin was first detected in cerebrospinal fluid (CSF) and brain homogenates some three decades ago (Margolis and Altszuler, 1967; Le Roith et al., 1983); however, it was only after the identification of the insulin receptor (IR) in the brain that the search for a role for insulin in neural physiology and pathophysiology was intensified (Havrankova et al., 1978a). Control of food intake and thermogenesis are the most studied central effects of insulin (Woods et al., 1979), although recently, the participation of insulin in the control of learning and memory (Zhao and Alkon, 2001) and in the development of degenerative diseases such as Alzheimer and Parkinson's, has been emphasized (Mattson et al., 1999). This chapter focuses on the effects of aging upon insulin action in the central nervous system (CNS), its possible implications in age-related feeding disturbances, and in the development of neurodegenerative diseases.

2. A synoptic view of the molecular mechanisms of insulin signaling

Insulin is produced by pancreatic β-cells in response to glucose and other metabolic, hormonal, and neural inputs (Ashcroft et al., 1994). It acts in target tissues by binding and activating a heterotetrameric transmembrane receptor that belongs to a subfamily of receptor tyrosine kinases. This subfamily includes the insulin-like growth factor (IGF)-1 receptor (IGF-1R), the insulin receptor-related receptor, and the IR itself (Patti and Kahn, 1998). Once bound to the α-extracellular subunit of the IR, insulin stimulates the catalytic activity of the β-transmembrane subunit (Kasuga et al., 1982) leading to transphosphorylation at tyrosine residues and initiation of an intracellular cascade that drives the signal toward specific subcellular sites and orchestrates cell responses (White and Kahn, 1994).

So far, at least nine intracellular direct substrates of the IR have been identified (Fig. 1). The insulin receptor substrate (IRS) family, which is composed of four members (IRS1-4), has been the most intensively studied (White, 2002). Additionally there are Shc (Sasaoka and Kobayashi, 2000), Gab1 (Rother et al., 1998), p60dok (Danielsen and Roth, 1996), Cbl (Ribon et al., 1998), and APS (Ahmed et al., 2000) proteins. Tyrosine-phosphorylated IR substrates act as docking sites for downstream proteins that contain SH2 domains. Some of these proteins are adapter molecules that induce the activation of small G proteins, while others are themselves enzymes. As a rule, at least four well-defined paths are activated by the insulin signal. The first one depends on IRS1/2 tyrosine phosphorylation, which leads to p85/PI3kinase binding and activation of its catalytic activity (Ruderman et al., 1990; Folli et al., 1992). The ensuing membrane accumulation of phosphatidylinositol$_{3,4,5}$ P3 (PI3P) provides binding sites for PDK and Akt. Through this pathway insulin participates in the control of glycogen, lipid, and protein synthesis (Mendez et al., 1997; Wada et al., 1999), apoptosis (Yenush et al., 1998), cell growth and differentiation (Conejo and Lorenzo, 2001), glucose metabolism (Sakaue et al., 1997), and GLUT4 migration toward the cell surface (Kanai et al., 1993; Kelly and

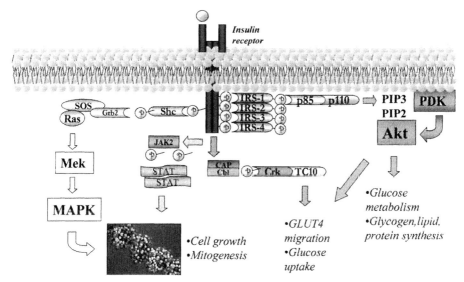

Fig. 1. Schematic illustration of the main molecular components of the insulin-signaling machinery. The activation of the heterotetrameric insulin receptor by the binding of insulin leads to engagement of several distinct pathways which control a number of functions on a given cell or tissue. Cell growth and mitogenic stimulus may be achieved through the activation of the MAP-kinase cascade or the JAK/STAT signaling pathway, which are depicted in the left-hand side of the figure. JAK/STAT signaling may be employed for cross-talk between distinct pathways, as well. Metabolic events are controlled by the PI3-kinase/Akt pathway and by the CAP/Cbl pathway, which are depicted in the right-hand side of the figure. The PI3-kinase pathway may provide survival and growth stimulus, as well.

Ruderman, 1993). Recent evidence strongly suggests that this latest function is achieved only if a second branch of the insulin-signaling cascade is properly activated (Baumann et al., 2000). This pathway involves the tyrosine phosphorylation of the Cbl protooncogene (Ribon and Saltiel, 1997) and leads to cytoskeleton adaptations required for adequate GLUT4 vesicle translocation (Chiang et al., 2001). A third branch of the insulin-signaling cascade leads to the activation of the mitogen-activated protein (MAP)-kinase cascade (Peyssonnaux and Eychene, 2001). This cascade is composed of a series of kinases that are sequentially stimulated and lead to phosphorylation of ERK, which once activated, migrates to the nucleus and initiates a transcriptional program (Davis, 1995). The MAP-kinase cascade may be activated via different proximal events, for instance following Shc tyrosine phosphorylation by the IR, following Grb2 recruitment by tyrosine-phosphorylated IRSs or even as a consequence of Gab1 engagement (Rother et al., 1998). The fourth pathway that can be activated by insulin is the JAK2/STAT signaling pathway (Saad et al., 1995; Velloso et al., 1998). This pathway is commonly engaged by cytokine-derived stimuli and delivers a rapid signal to the nucleus (Levy and Darnell, 2002). Recruitment of JAK2 by insulin is involved in molecular cross-talk between insulin and angiotensin II (Velloso et al., 1996) and may participate in coordinated control of food intake by acting as an interface between leptin and insulin signaling in the hypothalamus (Carvalheira et al., 2001).

3. Insulin signal transduction in the central nervous system

For insulin to promote any physiological response in the CNS, at least two requirements must be fulfilled. First, the hormone should be present at a physiological concentration in the CSF and brain tissue; secondly, IR and at least some of its substrates should coexist in certain areas of the brain. Havrankova et al. (1978b) provided the evidence for this first requirement. According to their studies, insulin concentrations in extracts of whole rat brain are higher than plasma insulin levels, and brain insulin is indistinguishable from pancreatic islet insulin, based on its immunological and biochemical characteristics. Initially, the same authors reported that levels of brain insulin suffered variation independently of peripheral levels of the hormone and therefore, proposed that insulin present in the brain could be, at least in part, synthesized in that region of the body (Havrankova et al., 1979). Hormone-binding studies and immunohistochemical characterization of IR in choroids plexus provided evidence that most, if not all, of the insulin in the brain is produced by β-cells and delivered to the CNS through a complex mechanism of receptor-mediated transport (Baskin et al., 1986; Frank et al., 1986; Pardridge, 1986). *In situ* hybridization and immunohistochemical studies have provided a complete mapping of the IR distribution in the CNS (Zahniser et al., 1984; Zhao et al., 1999; Zhao and Alkon, 2001), and the highest concentration of IR is found in choroid plexus, olfactory bulb, pyriform cortex, amygdaloid nucleus, hippocampus, hypothalamic nuclei, and cerebellar cortex (Marks et al., 1991) (Fig. 2). High levels of transcripts exist in the cerebral cortex, however, the resulting amount of protein is not proportional. It has been proposed that an accelerated rate of IR protein degradation might take place in this anatomic territory, which could explain the apparent discrepancy between mRNA and protein levels (Zhao and Alkon, 2001).

The rapidly increasing incidence of obesity in most regions of the world (Kopelman, 2000) has stimulated research on the mechanisms of hypothalamic control of food intake and energy expenditure. In parallel with leptin, insulin provides a potent anorexigenic signal to the brain (Woods et al., 1979), driven by a coordinated control of specific hypothalamic neurons. To characterize the main sites of insulin action in hypothalamus, double-immunostaining techniques and image acquisition by confocal microscopy have been employed. These studies have demonstrated that most hypothalamic IR expression occurs in the arcuate nucleus, a site known to coordinate the incoming anorexigenic signals from the periphery. Moreover, IR immunoreactivity is also detected in the periventricular nucleus and in some scattered neuron bodies in the posterior and lateral hypothalamus (Torsoni et al., 2003) (Fig. 3). According to these studies and some previous reports, most IR reactivity in the CNS is localized in neuron bodies (Schwartz et al., 1992; Torsoni et al., 2003), whilst only low levels of IR are present in cells of the glia (Marks et al., 1991).

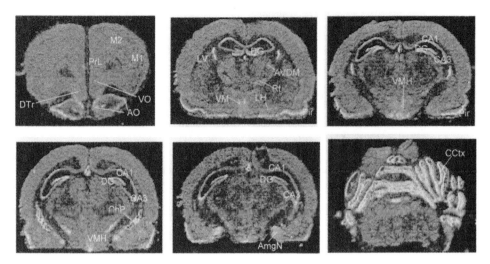

Fig. 2. Distribution of the insulin receptor mRNA in central nervous system. Adult rat brain sections were hybridized with an IR riboprobe labeled with [35]S. Highest insulin receptor mRNA is detected in cerebellar cortex (CCtx), hypothalamic nuclei [lateral hypothalamus (LH) and ventromedial nucleus of hypothalamus (VMH)], anterior olfactory nucleus (AO), and choroids plexus (Chp). PrL, prelimbic cortex; M1 and M2, primary and secondary motor cortexes, respectively; VO, ventral orbital cortex; DTr, dorsal transition zone; LV, lateral ventricle; AM, antero-medial thalamic nucleus; VA, ventral anterior thalamic nucleus; Rt, reticular thalamic nucleus; AVDM, anteroventral thalamic nucleus; VM, ventromedial thalamic nucleus; Pir, piriform cortex; CA1, CA3, and CA4, hippocampus; DG, dentate gyrus; AmgN, amygdaloid nucleus. [This figure was a kind donation of Dr. Zhao W.-Q., from the Blanchette Rockefeller Neurosciences Institute at the Johns Hopkins University (Zhao and Alkon, 2001).]

Since the IR is not widely distributed in the CNS, it has become clear that, in order to obtain a full understanding of the multiple actions of insulin in the brain, a meticulous characterization of the localization of the IR's main substrates and of the molecular events triggered by insulin activation in the brain, is necessary. Folli et al. (1994) demonstrated that IRS1 immunoreactivity is widely distributed in neurons localized in several areas of the brain and spinal cord. The cerebral cortex, the hippocampus, many hypothalamic and thalamic nuclei, the basal ganglia, the cerebellar cortex, the brainstem nuclei, and the lamina X of the spinal cord are particularly rich in immunopositive nerve cells. In these areas, most of the neurons immunoreactive for IRS1 are also stained by anti-IGF1 receptor antibodies and to a lesser extent by anti-IR antibodies. PI-3 kinase, a well-known participant in the insulin-signaling pathway, is also detected in several areas containing IR and IRS1. In a recent study, Torsoni and colleagues showed that most IR in the hypothalamus coincides with expression of IRS2 but not with IRS1, which is present mostly in posterior hypothalamus (Fig. 3). Hormone-induced receptor and substrate phosphorylation studies confirmed that IRS2 acts as the main substrate for IR in the hypothalamus (Torsoni et al., 2003).

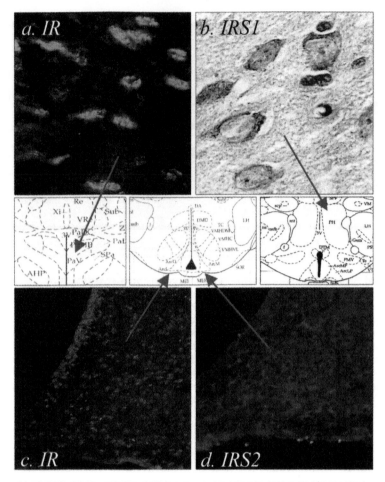

Fig. 3. Immunohistochemical evaluation of IR, IRS1, and IRS2 protein distribution in hypothalamus of rats. Paraformaldehyde-fixed rat hypothalamic sections were primarily incubated with anti-IR (a and c), -IRS1 (b), or -IRS2 (d) specific antibodies. Visualization of immunoreactivity was obtained by using secondary antibodies conjugated with FITC (a, c, and d) and photographed under a fluorescence microscope, or secondary biotinylated antibodies, recognized by avidin-conjugated peroxidase and photographed under an optical microscope (b). Approximate anatomic sites staining for each protein are depicted in the schemes and are indicated with arrows. Arc, arcuate nucleus; PaV, paraventricular nucleus; PH, posterior hypothalamic area; 3V, third ventricle. (From Torsoni et al., 2003.)

In a number of tissues that physiologically respond to insulin, Shc- and ERK-signaling pathways are engaged and provide robust mitogenic and growth-stimulating signals (Peyssonnaux and Eychene, 2001). Since elements that participate in these signaling cascades are widely distributed and participate as intermediaries in signaling events triggered by a wide spectrum of hormones, cytokines and growth factors, several groups have attempted to characterize their presence and functional spectrum in neural tissue. In the CNS, low levels of Shc are expressed in most areas; however, two homologs named N-Shc and Sck, are widely distributed and expressed at high protein levels in the brain. Apparently, these proteins may concentrate most of the Shc-like adapter function in this part of the body (Cattaneo and Pelicci, 1998; Nakamura et al., 1998). Nevertheless, although present at low levels in most parts of the CNS, Shc was detected in forebrain and cerebellar cortex homogenates, and was shown to have responded to acutely infused insulin undergoing tyrosine phosphorylation and association with Grb2 (de LA Fernandes et al., 1999). Conversely, brain-specific homologs of proteins that belong to the MAP-kinase cascade do not seem to play a major role in the activation of this pathway in the CNS. ERK1 [M(r) 44 kDa] and ERK2 [M(r) 42 kDa], the two major extracellular signal-regulated protein kinases, are present in different regions of rat brain. The aggregate ERK concentration (ERK1 plus ERK2) is relatively high throughout the brain, ranging from lowest levels in the cerebellum to highest levels in the nucleus accumbens. The ratios of ERK1 protein to ERK2 protein vary along a rostral–caudal gradient from a low of 0.16 in frontal cortex to a high of 1.5 in pons and medulla. Subcellular fractionation studies reveal that both ERK1 and ERK2 are enriched in the synaptosomal and cytosolic fractions, whereas ERK2 is also enriched in the microsomal fraction (Ortiz et al., 1995).

Proteins that participate in signal transduction through the JAK/STAT signaling pathway have also been studied in the brain. The presence and functional status of JAK family members was initially evaluated in immortalized striatum-derived nestin-positive adult and embryonic cell lines. In both cases JAK2 was detected and shown to respond to IL-3 with tyrosine phosphorylation and participation in the induction of proliferative response (Cattaneo et al., 1996). In brain of living rats the presence and molecular response of JAK2, STAT1, and STAT3 was detected in ciliary neurotrophic factor, EGF and FGF responsive neurons, suggesting a role for this signaling pathway in neural growth and mitogenesis (De-Fraja et al., 2000). Finally, in a recent study, insulin- and leptin-responsive neurons in the hypothalamus were shown to express JAK2 and STAT3 (Carvalheira et al., 2001). In this anatomic area, JAK/STAT signal transduction seems to participate in control of energy expenditure and food intake (Tartaglia et al., 1995; Carvalheira et al., 2001).

The progressive accumulation of data on tissue distribution of proteins that participate in insulin signal transduction events in the CNS-impelled research groups to characterize the region-specific pattern of insulin-induced molecular activation of the insulin-signaling pathway in the brain. In experimental animals, both peripheral and intra-cerebro-ventricular (ICV) injection of exogenous insulin promotes tyrosine phosphorylation of the IR in various areas of the brain. Infusion of insulin in the cava vein leads to tyrosine phosphorylation of the IR present in brain protein extracts after 10–15 min. This event is shortly followed by tyrosine phosphorylation of IRS2 and Shc and to a lesser extent of IRS1 (de LA Fernandes et al., 1999; Torsoni et al., 2003). There are clear regional differences in the pattern of activation of signal transduction molecules induced by insulin. Molecular activation signals are easily detected in hypothalamus and cerebellum (de LA Fernandes et al., 1999; Carvalheira et al., 2001; Torsoni et al., 2003), whilst in cerebral cortex the magnitude of events is smaller (de LA Fernandes et al., 1999). These regional differences seem to be related to distribution and rate of turnover of IR and IGF-1R, and to specific usage of intermediary substrates by insulin or IGF1 primary signals (Folli et al., 1994).

In conclusion, proteins that participate in insulin signal transduction in peripheral tissues are present in several areas of the CNS and respond, to a certain degree, to peripheral or centrally administered insulin. This fact strongly suggests that insulin plays physiologic roles in the brain and may participate in the control of functions extending from the regulation of food ingestion to cellular growth and fine behavioral and learning responses.

4. Effects of aging on insulin signal transduction in peripheral tissues

Aging leads to progressive deterioration of several functions controlled by insulin (Fink et al., 1983; Goodman et al., 1983). Patients with diabetes mellitus tend to present a worsening of the laboratorial parameters used to evaluate metabolic control (Fink et al., 1983), and even nondiabetic individuals present some degree of glucose intolerance or signs of insulin resistance during aging (Barzilai and Gupta, 1999).

In experimental animals, insulin resistance that develops during aging is associated with glucose intolerance as determined by euglycemic–hyperinsulinemic clamp (Nishimura et al., 1988). This phenomenon is accompanied by impaired insulin-induced glycogen synthesis and inhibition of glucose output by the liver (Goodman et al., 1983; Matthaei et al., 1990).

Nadiv et al. (1992) demonstrated that 2-year-old female Wistar rats present a 50% impairment of in vivo IR autophosphorylation in comparison to 3-month-old rats. They also showed that in vitro kinase activity of IR of old rats were preserved. In a subsequent study, the same group demonstrated that livers of old rats present an elevated activity of protein tyrosine phosphatase and an increased plasma membrane viscosity (Nadiv et al., 1994). It was then suggested that increased dephosphorylation of the IR β-subunit plus changes in the lipid composition of cell membrane contributed to reduce the activity of IR and finally to impair insulin signaling in the liver. In recent years, a series of studies were performed with the objective of further characterizing the molecular events that might be related to aging-dependent insulin resistance. Carvalho et al. (1996) demonstrated that aging induces no significant changes of IR protein amount in skeletal muscle and liver of rats. On the other hand, the IRS1 protein concentration and insulin-induced tyrosine phosphorylation of the IR and IRS1 are severely reduced in the liver and muscle of senile rats. Moreover, insulin-induced IRS1-associated PI3-kinase activity is also impaired in liver and muscle of senile animals. Paez-Espinosa et al. (1999) reported that aging does not modulate the amount of Shc and Grb2 in liver, muscle, and adipose tissue of rats. However, following an acute insulin stimulus, tyrosine phosphorylation of Shc and Shc–Grb2 association are significantly increased in all of the aforementioned tissues of the aging rodents.

In a recent study, Zecchin et al. (2003) observed that obese middle-aged rats demonstrate marked insulin resistance, which parallels the reduced effects of this hormone in the insulin-signaling cascade in muscle. The PI3-kinase/Akt pathway is preserved in aorta of these rats, leading to a normal activation of endothelial nitric oxide synthase. In contrast, in spontaneously hypertensive rats, this pathway is severely inhibited with reduction in eNOS protein concentration and activation. Both animals, however, present higher concentrations and higher tyrosine phosphorylation of MAP kinase isoforms in aorta.

An involvement of free-radical accumulation and a nonenzymatic reaction between sugars, lipids, and proteins has been postulated to occur along the course of aging and diabetes mellitus. Based on this hypothesis, structurally and functionally altered proteins cause changes in the cell's signal trasduction process. A correlation of the higher expression of advanced glycation end products (AGEs) and advanced lipoxidation end products (ALEs) with disturbed cell and tissue function provide evidence to support this hypothesis (Vlassara et al., 1992, 1994). This process involves the degradation of the inhibitory signaling protein IkB, and allows NF-kB to migrate to the nucleus to promote pathological transcription of pro-inflammatory factors such as TNF-α and IL-1β, in addition to vascular cell-signaling defects mediated by activation of PKC (Evans et al., 2002). This long-term attack to the organs responsible for controlling organism function and systemic integration, such as the brain, the liver, and blood vessels, lead to the disruption of the equilibrium and peripheral organ deterioration (Sell and Monnier, 1990; Baynes and Thorpe, 2000).

It has been hypothesized that during life, AGEs compromise cell signaling as indicated by their effects upon the EGF receptor *in vitro* (Portero-Otin et al., 2002). The correlation between AGEs accumulation and insulin signaling remains to be explored; however, a recent study has indicated that aberrant N-linked glycation of the precursor molecule of IR determines its degradation before initiating activity in 3T3-L1 adipocytes (Hwang et al., 2000).

Aging also affects the physiology of exocrine glands that provide the lubricant, nourishing, and protective secretions that compose saliva and tears. Dry eyes and mouth are a consequence of this dysfunction (Sreebny and Valdini, 1988; Schein et al., 1997). Recent studies have demonstrated that hormones have an important function in the correct activity of salivary and lacrimal glands (Sullivan et al., 1998). Exocrine glands, which are part of the immune secretory system, are tightly regulated by neural inputs and undergo dramatic impairment of their functional activity during senility. In aging rats, the total amount of protein-secreted by lacrimal glands, and the lacrimal gland/bodyweight ratio is reduced (Bromberg and Welch, 1985), while the deposition of connective tissue and inflammatory cell infiltration is increased (Damato et al., 1984; Sullivan et al., 1990).

Insulin is one of the hormones, which plays a key role in lacrimal and salivary tissues as evidenced by its action in acinum cell growth *in vitro*, and by the dysfunction observed in animal models and humans with diabetes mellitus

(Lindberg et al., 1993; Pellegrini et al., 1997; Dogru et al., 2001). Previous studies have shown that insulin signaling is impaired in lacrimal and salivary glands of old rats. Aging reduces IR phosphorylation in both exocrine glands and extends downstream defective signaling to STAT-1 phosphorylation, which may be responsible in part for the mechanisms of senescence-related dysfunction of tissues (Rocha et al., 2002a).

The eyes are a straight continuation from the CNS in developmental, functional, and metabolic aspects. Insulin signaling appears to be crucial for retina development and axonal linkage to the CNS (Song et al., 2003). In addition, recent work has shown that light is able to stimulate PI3-kinase, through activation of IR in mice, in a process thought to promote retina neuroprotection (Rajala et al., 2002). Aging, and diabetes mellitus are known to induce damage to many of the eye components, leading to retinopathy, cataracts, and dry eyes. Recent studies have characterized the presence and role for insulin in this area of the body. Accordingly, insulin, IR, and all molecular elements needed for the proper activity of the insulin signaling apparatus are present and functionally operative in most of the eye tissues evaluated (Rocha et al., 2000, 2002b,c). Aging and anomalous insulin signaling in these tissues may lead to a number of functional defects in the lens, retina, and conjunctiva (Civil et al., 2000; Zatechka and Lou, 2002).

As a whole, aging promotes a stepwise and tissue-specific modulation of different branches of the insulin-signaling cascade, impairing metabolic-related functions earlier on in life than growth-related functions, which may contribute to some aging-related proliferative disorders, such as occurs in vascular diseases.

5. Central nervous system mechanisms that control aging

A series of observational data collected since the 1960s (Pecile et al., 1965; Singh and Bachhawat, 1965; Mattson, 2002; Mattson et al., 2002) provide evidence that the CNS participates in the control of life span. Part of this control is achieved through regulation of physiological processes that are essential for life, such as feeding, thermogenesis, blood pressure, cardiac and respiratory rate, and stress response. Moreover, the specialized capacity of the brain for learning, accumulating experiences, and using accumulated learning to solve problems provides efficient means for increasing life span, which is exemplified by the fact that the life expectancy of human beings has doubled during the last century.

In addition to the mechanisms cited above, which may be considered indirect means for controlling longevity, recent studies demonstrate that participation of brain as a key regulator of life span may depend on its capacity for modulating signaling pathways that control metabolism and oxidative stress (Tower, 2000; Tatar et al., 2003).

In the fly *D. melanogaster*, whole-body mutations that increase the catalytic activity of Mn-SOD and Cu/Zn-SOD, two antioxidant enzymes that catalyze the dismutation of reactive oxygen species significantly increase the life span of the insect (Tatar et al., 2003). Similarly, upregulation of the chaperone protein, HSP-70, may extend the life span of *D. melanogaster* by up to 20% (Tatar et al., 1997). If Cu/Zn-SOD activity is upregulated exclusively in neurons, the lifetime of *D. melanogaster* is increased as well (Parkes et al., 1998). Moreover, treatment of mice with the antioxidant compound, melatonin, prolongs the life span of experimental animals by up to 20% (Oxenkrug et al., 2001). Thus, increasing antioxidant efficiency in the whole body or in the brain exerts a remarkable effect upon the life span of experimental animals.

In the nematode *Caenorhabditis elegans*, mutation of the gene *daf-2*, which encodes a homolog of the IR, and of the gene *age-1*, which encodes a homolog of PI3-kinase, leads to increased longevity (Kenyon et al., 1993; Lin et al., 2001). Similarly, in *D. melanogaster*, defects in *chico* (which encodes a homologue of the IR) promote a significant increase in life span (Clancy et al., 2001; Tatar et al., 2001). In mammals, activation of CNS insulin and IGF-1-signaling pathways induce pituitary production of GH and TSH, which in turn promote the peripheral activation of a series of metabolic pathways that favor aging (Tatar et al., 2003). Mutations of the genes *Prop1* and *Pit1* decrease GH, TSH, and prolactin production, resulting in reduced body size and up to 60% increase in life span (Brown-Borg et al., 1996; Flurkey et al., 2001). Similar phenotypic characteristics are achieved with the knockout of the GH receptor (Sims et al., 2000). In such mutants, small body size and increased longevity are accompanied by reduced serum IGF-1 and insulin, which are paralleled by anomalous pancreatic islet development (Dominici et al., 2002). The most remarkable evidence of participation of the IGF-1-signaling pathway in the control of aging was obtained with the disruption of the IGF-1 receptor gene (Holzenberger et al., 2003). Female *igf-1r*$^{+/-}$ have normal body size, sexual maturation, and metabolic rate; however, they live up to 30% longer than wild-type mice.

Food restriction increases life span by up to 40% in rodents (Weindruch and Sohal, 1997). Hypoinsulinemia is a major consequence of food restriction and has been suggested to be an important participant of the mechanisms that link low caloric intake to extended lifetime (Mattson et al., 2002). Hypoinsulinemic animals exhibit increased insulin sensitivity and enhanced glucose tolerance (Kalant et al., 1988), which are accompanied by the improvement of several metabolic parameters. Moreover, food restriction increases resistance to development of neurodegenerative diseases, which certainly contributes to extended life span (Bruce-Keller et al., 1999; Mattson et al., 1999).

The regulation of expression of genes that participate in cell survival routes has been detected in animals submitted to low caloric intake, and these effects may contribute to increasing life span (Mattson et al., 2002). Brain-derived neurotrophic factor (BDNF) expression is induced in the brain of calorie-restrained rodents (Lee et al., 2000). BDNF signals through a receptor of the tyrosine-kinase receptor family and activates PI3-kinase and Akt signaling (Foulstone et al., 1999). Acting through this pathway, BDNF induces insulin-dependent-like mechanisms of cell survival in the CNS (Foulstone et al., 1999). Recent evidence suggests that activation of insulin-like BDNF signaling enhances learning and memory (Minichiello et al., 1999), and rescues neurons damaged following metabolic and oxidative stress

(Gary and Mattson, 2001). Moreover, reduced expression of BDNF leads to obesity, whilst central administration of the peptide reduces bodyweight and improves glucose metabolism in diabetic, leptin receptor-deficient mice (Tonra et al., 1999).

Ciliary neurotrophic factor (CNTF) is a cytokine-like neurotrophic factor that was primarily characterized for its neuron-survival-inducing properties (Inoue et al., 1996). The infusion of CNTF in experimental animals leads to severe anorexia and development of cachexia (Henderson et al., 1994). Some of its actions are comparable to those induced by IL-6, especially the induction of the acute-phase response by the liver (Espat et al., 1996). However, in contrast to IL-6, exogenous CNTF administration in rodents induces severe lean-mass wastage and anorexia (Fantuzzi et al., 1995; Espat et al., 1996), a phenomenon driven by centrally mediated mechanisms (Espat et al., 1996; Martin et al., 1996). The characterization of the leptin receptor, ObR, by Tartaglia et al. (1995), revealed a striking similarity with the CNTF receptor (Davis et al., 1991) and boosted interest in the anorexigenic properties of the neurotrophic factor. In a recent study, Lambert et al. (2001) revealed that CNTF acts through receptors localized in the arcuate nucleus activating an intracellular pathway similar to that utilized by leptin. Interestingly, differently to leptin, the administration of CNTF to diet-induced obese mice promotes weight loss, which is maintained after the interruption of the treatment. Moreover, metabolic parameters of glucose-intolerant animals can be reverted by CNTF. Thus, this cytokine-like neurotrophic factor, acting through the same intracellular signaling pathway utilized by leptin, is capable of inducing metabolic adjustments that favor a prolonged life span.

Figure 4 summarizes the integration of the CNS signals that act through insulin/ IGF-1-signaling pathways to control several physiological parameters that affect life span. In the left-hand side of the figure, are the mechanisms that protect an individual from central and peripheral insulin resistance possibly providing a tight control against pro-inflammatory and pro-oxidative events. BDNF and CNTF may act through, as yet unknown, physiological mechanisms to activate protective pathways. In the right-hand side, are depicted the mechanisms that induce insulin resistance and favor pro-aging events.

6. Effects of aging on insulin signal transduction in brain

The accumulated evidence suggesting that insulin and IGF-1 signaling might play important roles in the process of aging has raised fundamental questions regarding the effect of senescence upon the expression and functional activation of proteins that participate in insulin/IGF-1 signaling in the brain. Early studies have attempted to characterize the concentration of insulin and IGF-1 in the CNS. Thus, the transcription of IGF-1 in brain was found to be highest during embryonic life falling progressively thereafter, until reaching a steady state during adulthood (Rotwein et al., 1988). From adult life to senescence no further fall is detected (Sonntag et al., 1999). In contrast, the concentration of immunoreactive IGF-1 in brain extracts was shown to fall progressively from middle adult age to senility,

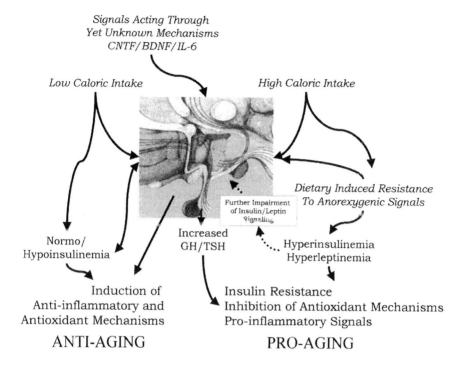

Fig. 4. Schematic illustration of some events that contribute for central control of longevity. In the left-hand side of the figure are depicted events that ultimately lead to hypoinsulinemia and favor longevity. In the right-hand side of the figure are depicted events that lead to hyperinsulinemia/hyperleptinemia, and activate pro-aging mechanisms.

suggesting that defective local translation, or impaired transport of IGF-1 through the blood–brain barrier, might contribute to the differences observed (Sonntag et al., 1999).

Since the local production of insulin in the CNS does not seem to play a major role in the actions of this hormone in the brain, variations of insulin concentration in this territory are dependent on serum insulin concentration and, particularly, on the kinetics of the blood–brain barrier transport system. Serum insulin levels are particularly high at birth falling rapidly from day 1 to day 21 of life, and then remaining steady until middle adult life. Thereafter, insulin levels rise progressively until senility (Fernandes et al., 2001). Since the earliest reports relating the presence of insulin in the brain (Havrankova et al., 1978b), researchers have noticed that no straight correlation exists between serum and brain or CSF insulin (Havrankova et al., 1978b). During aging, a dichotomy between serum insulin levels and brain insulin levels further supports this notion. Thus, although not an established concept, most studies suggest that in contrast to serum insulin the levels of this hormone in brain fall progressively during aging (Bernstein et al., 1984; Frolich et al., 1998). This fact has raised the possibility that progressive, age-dependent impairment of the insulin-transporting apparatus at the blood–brain barrier might play a central role in the control of insulin concentration in the brain.

The receptors for insulin and IGF-1 also present some age-dependent regulation (Sonntag et al., 1999). High protein levels of the IR are present in rat brain during embryonic life. However, following birth a subtle decrease in receptor amounts is observed in several areas of the brain. In contrast, in the forebrain cortex, cerebellum and some hypothalamic nuclei, high concentrations of IR persist during adulthood and senility (de LA Fernandes et al., 1999; Fernandes et al., 2001; Torsoni et al., 2003). The IGF-1 receptor is also expressed at high amounts in several areas of the brain. In the cerebellum and other subregions of the forebrain the binding of labeled IGF-1 to brain slices increases slightly from birth to middle adulthood and then decreases progressively (Pomerance et al., 1988). A few studies have evaluated the distribution and amount of insulin and IGF-1/IGF-2-binding sites in the brain of aging mammalians (Kar et al., 1993; Dore et al., 1997). In all studies, higher concentrations of insulin and IGF-binding sites in the brain are observed during embryonic and early postnatal life (Pomerance et al., 1988; Marks and Eastman, 1990; Potau et al., 1991). From birth to adulthood a progressive fall in binding sites occurs in most regions of the CNS (Marks and Eastman, 1990). Consequently, from adult life into senescence the concentration of binding sites remains steady (Marks and Eastman, 1990). Most binding occurs in the olfactory bulb, cortex, hippocampus, choroid plexus, and cerebellum. Although the levels of insulin and IGF-1 receptors in several areas of the brain do not change during aging (Dore et al., 1997; Fernandes et al., 2001), the number and affinity of high-affinity binding sites for insulin are modulated with aging (Zaia and Piantanelli, 2000). This occurrence does not seem to influence the total binding affinity of either insulin or IGF-1 in the CNS (Dore et al., 1997).

Following ligand binding, IR undergoes autophosphorylation and, thereafter recruits intracellular substrates that transduce the insulin signal toward specific subcellular compartments (Saltiel and Kahn, 2001). Some studies have evaluated insulin and IGF-1-signaling events in the brain of aging experimental animals. Following an acute infusion of insulin, a clear increase in IR tyrosine phosphorylation is observed in the hypothalamus and cerebellar cortex (de LA Fernandes et al., 1999; Torsoni et al., 2003). However, this phenomenon is less evident in cerebral cortex (de LA Fernandes et al., 1999). During aging, a significant decrease in insulin-induced tyrosine phosphorylation of the IR is observed in the forebrain, cerebellum cortex, and hypothalamus, suggesting that during senescence the molecular activation of the first step of insulin signaling is impaired (Fernandes et al., 2001). With regard to the regulation of IGF-1 receptor function during development and aging the picture is not so clear. According to Girault et al. (1992), during the first steps of extra-uterine life a significant increase in tyrosine phosphorylated proteins, including the IGF-1 receptor, is observed in the brain of mice. During adulthood, the magnitude of these signaling events is maintained steady (Girault et al., 1992).

In peripheral organs that act as targets for insulin action such as liver, skeletal muscle, and adipose tissue, the tyrosine phosphorylation of IRS1 and IRS2 following an acute infusion of insulin is a clear and easily reproducible phenomenon (Kahn, 1994; Saad, 1994; White and Kahn, 1994). In the brain, however, the induction of tyrosine phosphorylation of IRS1 by acutely infused insulin is not a ubiquitous event (Folli et al., 1994; de LA Fernandes et al., 1999; Torsoni et al., 2003) and apparently occurs only in some scattered neurons (Torsoni et al., 2003). This fact has contributed to hamper efforts to characterize age-dependent modulation of insulin signaling toward IRSs in the CNS. Similarly, although several studies have demonstrated a prominent role for brain PI3-kinase and Akt in the control of central and peripheral metabolism and in the regulation of cell survival (Foulstone et al., 1999; Cheng et al., 2000; Yamaguchi et al., 2001), no objective studies have evaluated the role of aging upon expression and activity of these enzymes in the CNS.

Another important substrate of the IR, Shc, is present in the CNS and is responsive to insulin undergoing tyrosine phosphorylation (de LA Fernandes et al., 1999). The protein levels of Shc in the forebrain and cerebellum do not modify significantly from birth to senility (de LA Fernandes et al., 1999). However, in both regions basal and insulin-induced tyrosine phosphorylation of Shc decrease abruptly from early extra-uterine life to day 21 of life. Thereafter, until senility, no further decrease is detected (de LA Fernandes et al., 1999).

Following tyrosine phosphorylation, Shc binds to the 23 kDa adapter protein, Grb2, which is constitutively associated with the proline-rich domain of SOS (son-of-sevenless), a guanine nucleotide-exchange factor for p21-Ras. Once stimulated, the Shc/Grb2/SOS complex drives the signal toward activation of the MAP kinase-signaling cascade and consequently modulates gene transcription and mitogenesis (Peyssonnaux and Eychene, 2001). Aging does not affect the amount of Grb2 protein expression in forebrain and cerebellum (Fernandes et al., 2001), although basal and insulin-stimulated Shc/Grb2 binding undergoes a significant fall from day 1 to day 21 of extra-uterine life, remaining steady thereafter (Fernandes et al., 2001).

Multiple mechanisms participate in the termination of signaling events. Shutting down a given intracellular signaling pathway is essential for avoiding overstimulation and for preparing the cell or tissue for an oncoming stimulus. Signaling pathways that utilize tyrosine phosphorylation of substrates as a mechanism for transducing signals may be controlled by several mechanisms including internalization and degradation of the transmembrane receptor, activation of phosphatases, serine or threonine phosphorylation of substrates, and direct interaction of inhibiting proteins such as proteins of the SOCS (suppressor of cytokine signaling) or PIAS (protein inhibitor of activated STATs) families (Hilton, 1999; Kile et al., 2001). Insulin/IGF-1 and leptin signaling may be controlled by tyrosine phosphatases or by proteins of the SOCS family, specifically SOCS3 (Bjorbaek et al., 1998, 1999; Emanuelli et al., 2000; Rui et al., 2002). Two recent studies have evaluated the effect of aging upon the expression of SHP2, a tyrosine phosphatase with dual – (activating and inhibiting) activity (Fernandes et al., 2001), and SOCS3 (Peralta et al., 2002). In the first study, Fernandes et al. provided evidence that SHP2 expression in brain decreases significantly from early postnatal life to adulthood, remaining steady thereafter, until senility. In a second study, Peralta et al. (2002) demonstrated that SOCS3 expression increases progressively in the hypothalamus of aging rats, suggesting that age-dependent impairment of leptin and insulin signaling in this territory may be, at least in part, due to SOCS3 engagement.

Thus, although several lines of evidence suggest that central control of signaling pathways that modulate brain and peripheral metabolism participate actively in the control of aging, research characterizing the effects of aging on the expression, tissue distribution, and activity of proteins that participate in these signaling events is lacking. Figure 5 schematically summarizes the data concerning expression and functional status of several proteins involved in insulin/IGF-1 signaling in the brain of aging mammalians.

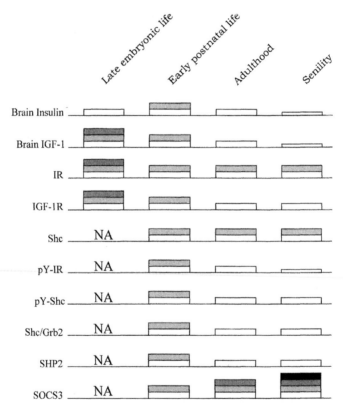

Fig. 5. Relative amounts of hormone/growth factors and molecular elements of the insulin/IGF-1-signaling pathway in brain during aging. Insulin and IGF-1 concentrations in brain extracts, protein amounts of insulin receptor (IR); IGF-1 receptor (IGF-1R); Shc; SHP2 and SOCS3, insulin-induced tyrosine phosphorylation of insulin receptor (pY-IR) and Shc (pY-Shc), and insulin-induced association between Shc and Grb2 (Shc/Grb2) were plotted relative to respective quantification during early postnatal life. Based on Bernstein et al., 1984; Pomerance et al., 1988; Rotwein et al., 1988; Marks and Eastman, 1990; Potau et al., 1991; Girault et al., 1992; Frolich et al., 1998; de LA Fernandes et al., 1999; Sonntag et al., 1999; Fernandes et al., 2001; Peralta et al., 2002.

7. Insulin signaling and neural functions that are affected by aging

Aging gradually compromises several functions controlled by the brain. Progressive loss of memory is one of the most common complaints in senile patients (Hanninen et al., 1994), and severe deterioration of memory and other cognitive functions are components of the clinical traits of Alzheimer's disease (Hanninen et al., 1994). Motor disabilities are also common manifestations of elderly patients, and severe damage of nigro-striatum neurons leads to Parkinson's disease (Nussbaum and Ellis, 2003). A modification of feeding daily rhythm and food quality choice, which may lead to age-related weight gain and body fat redistribution is also a phenomenon frequently observed in aging humans and animals (Rossner, 2001). Finally, defective immune responses, which augment the chances of developing tumors or acquiring infectious diseases, are also more evident in elder patients than in younger patients and animals (Ershler, 1993; Pawelec et al., 2001). In the face of so many clinical features that are common to aging patients, an obligatory question has been raised: are there any common links between all these phenomena?

Unfortunately, we are still far from answering this intriguing question. In recent years, though, a growing body of evidence has placed insulin and other central controllers of brain and whole-body metabolism into a pivotal position in this complex scenario.

Some recent studies have shown that patients with Alzheimer and Parkinson's diseases present some clinical and laboratorial characteristics of glucose intolerance and impaired insulin action (Hoyer, 1991; Mytilineou et al., 1994; Blass et al., 1997). The aberrant proteolysis of amyloid precursor protein (APP) seems to be implicated in the pathogenesis of Alzheimer's disease (Mattson, 1997; Mattson and Furukawa, 1997), and mice that are made mutants for APP develop an Alzheimer-like syndrome, as well as glucose intolerance, providing further support for an association between this age-related neurodegenerative disorder and defective insulin signaling (Mattson et al., 1999; Pedersen et al., 1999). To date, however, there is no definitive evidence that a metabolic disturbance precedes or participates in the pathogenesis of these neural disorders (Mattson et al., 1999).

Recent studies have provided further support for the participation of insulin-related signaling pathways in the pathogenesis of Alzheimer and Parkinson's disease. PI3-kinase and Akt intracellular signaling were shown to participate in the control of neuron survival through the functional modulation of Bcl-2 (Seo et al., 2002), and via participation in the cellular mechanisms that protect neurons from oxidative damage induced by $\alpha\beta$ amyloid (Martin et al., 2001). Moreover, a PI3-kinase-independent mechanism leading to increased activation of the MAP kinase cascade in peripheral tissues was detected in patients with a familial form of Alzheimer's disease (Zhao et al., 2002).

The molecular events that participate in age-related disturbances of feeding and metabolism are not clear. Aging leads to progressive hyperinsulinemia and possibly to hypothalamic resistance to insulin. Since insulin signaling to hypothalamus provides potent anorexigenic and thermogenic signals (Woods et al., 1979; Torsoni et al., 2003), resistance to its action would impair energy expenditure and increase the threshold for pro-feeding inputs. Moreover, due to cross-talk mechanisms, hypothalamic insulin resistance may modulate the robust leptin signal and further impair the central control of feeding and thermogenesis (Carvalheira et al., 2001). Thus, age-dependent defective insulin and leptin signaling in the hypothalamus may play an important role in feeding and energy expenditure disturbances observed during senescence.

Finally, senescence is also associated with the development of several diseases that are a direct or indirect result of an impaired immune response. Increased incidence of cancer in elderly patients is certainly the most recognizable consequence of defective immune surveillance. However, aging patients and animals are also prone to developing infectious diseases, which may have devastating consequences. Recent data have provided strong evidence of the participation of leptin in the control of lymphocyte function (Lord et al., 1998) and in the central control of immune response (Howard et al., 1999). According to Steinman et al. (2003) leptin may have a dual action upon the immune system, on one hand controlling the production of intermediary hypothalamic neurotransmitters such as CRH and MSH, which drive indirect actions in the Th1 response, whilst on the other hand exerting direct effects on immature lymphocytes. Thus, leptin seems to act as a robust linkage between nutrition and metabolic status and the immune system. Defective leptin signaling that develops with aging may participate in aging-related immunodeficiency.

8. Concluding remarks

The insulin/IGF-1-signaling pathway plays a central role in the control of feeding behavior, energy expenditure, metabolism, and cell growth and survival. By direct and indirect means, insulin and other signal molecules that employ this pathway participate in the control of aging. Current concepts suggest that, on one hand, aging controls insulin signaling by promoting a modulation of protein expression and function in several tissues including the CNS, whilst on the other hand, insulin signaling participates actively in the processes that control the aging rate. Further progress in the characterization of insulin signaling in the brain may offer new targets for therapy of diseases commonly associated with aging.

References

Ahmed, Z., Smith, B.J., Pillay, T.S., 2000. The APS adapter protein couples the insulin receptor to the phosphorylation of c-Cbl and facilitates ligand-stimulated ubiquitination of the insulin receptor. FEBS Lett. 475, 31–34.

Ashcroft, F.M., Proks, P., Smith, P.A., Ammala, C., Bokvist, K., Rorsman, P., 1994. Stimulus-secretion coupling in pancreatic beta cells. J. Cell Biochem. 55, 54–65.

Barzilai, N., Gupta, G., 1999. Interaction between aging and syndrome X: new insights on the pathophysiology of fat distribution. Ann. N.Y. Acad. Sci. 892, 58–72.

Baskin, D.G., Brewitt, B., Davidson, D.A., Corp, E., Paquette, T., Figlewicz, D.P., Lewellen, T.K., Graham, M.K., Woods, S.G., Dorsa, D.M., 1986. Quantitative autoradiographic evidence for insulin receptors in the choroid plexus of the rat brain. Diabetes 35, 246–249.

Baumann, C.A., Ribon, V., Kanzaki, M., Thurmond, D.C., Mora, S., Shigematsu, S., Bickel, P.E., Pessin, J.E., Saltiel, A.R., 2000. CAP defines a second signalling pathway required for insulin-stimulated glucose transport. Nature 407, 202–207.

Baynes, J.W., Thorpe, S.R., 2000. Glycoxidation and lipoxidation in atherogenesis. Free Radic. Biol. Med. 28, 1708–1716.

Bell, G.I., Kayano, T., Buse, J.B., Burant, C.F., Takeda, J., Lin, D., Fukumoto, H., Seino, S., 1990. Molecular biology of mammalian glucose transporters. Diabetes Care 13, 198–208.

Bernstein, H.G., Dorn, A., Reiser, M., Ziegler, M., 1984. Cerebral insulin-like immunoreactivity in rats and mice. Drastic decline during postnatal ontogenesis. Acta Histochem. 74, 33–36.

Bjorbaek, C., El-Haschimi, K., Frantz, J.D., Flier, J.S., 1999. The role of SOCS-3 in leptin signaling and leptin resistance. J. Biol. Chem. 274, 30059–30065.

Bjorbaek, C., Elmquist, J.K., Frantz, J.D., Shoelson, S.E., Flier, J.S., 1998. Identification of SOCS-3 as a potential mediator of central leptin resistance. Mol. Cell 1, 619–625.

Blass, J.P., Sheu, K.F., Piacentini, S., Sorbi, S., 1997. Inherent abnormalities in oxidative metabolism in Alzheimer's disease: interaction with vascular abnormalities. Ann. N.Y. Acad. Sci. 826, 382–385.

Bromberg, B.B., Welch, M.H., 1985. Lacrimal protein secretion: comparison of young and old rats. Exp. Eye Res. 40, 313–320.

Brown-Borg, H.M., Borg, K.E., Meliska, C.J., Bartke, A., 1996. Dwarf mice and the ageing process. Nature 384, 33.

Bruce-Keller, A.J., Umberger, G., McFall, R., Mattson, M.P., 1999. Food restriction reduces brain damage and improves behavioral outcome following excitotoxic and metabolic insults. Ann. Neurol. 45, 8–15.

Carvalheira, J.B., Siloto, R.M., Ignacchitti, I., Brenelli, S.L., Carvalho, C.R., Leite, A., Velloso, L.A., Gontijo, J.A., Saad, M.J., 2001. Insulin modulates leptin-induced STAT3 activation in rat hypothalamus. FEBS Lett. 500, 119–124.

Carvalho, C.R., Brenelli, S.L., Silva, A.C., Nunes, A.L., Velloso, L.A., Saad, M.J., 1996. Effect of aging on insulin receptor, insulin receptor substrate-1, and phosphatidylinositol 3-kinase in liver and muscle of rats. Endocrinology 137, 151–159.

Cattaneo, E., De Fraja, C., Conti, L., Reinach, B., Bolis, L., Govoni, S., Liboi, E., 1996. Activation of the JAK/STAT pathway leads to proliferation of ST14A central nervous system progenitor cells. J. Biol. Chem. 271, 23374–23379.

Cattaneo, E., Pelicci, P.G., 1998. Emerging roles for SH2/PTB-containing Shc adaptor proteins in the developing mammalian brain. Trends Neurosci. 21, 476–481.

Cheng, C.M., Reinhardt, R.R., Lee, W.H., Joncas, G., Patel, S.C., Bondy, C.A., 2000. Insulin-like growth factor 1 regulates developing brain glucose metabolism. Proc. Natl. Acad. Sci. USA 97, 10236–10241.

Chiang, S.H., Baumann, C.A., Kanzaki, M., Thurmond, D.C., Watson, R.T., Neudauer, C.L., Macara, I.G., Pessin, J.E., Saltiel, A.R., 2001. Insulin-stimulated GLUT4 translocation requires the CAP-dependent activation of TC10. Nature 410, 944–948.

Civil, A., van Genesen, S.T., Klok, E.J., Lubsen, N.H., 2000. Insulin and IGF-I affect the protein composition of the lens fibre cell with possible consequences for cataract. Exp. Eye Res. 70, 785–794.

Clancy, D.J., Gems, D., Harshman, L.G., Oldham, S., Stocker, H., Hafen, E., Leevers, S.J., Partridge, L., 2001. Extension of life-span by loss of CHICO, a Drosophila insulin receptor substrate protein. Science 292, 104–106.

Conejo, R., Lorenzo, M., 2001. Insulin signaling leading to proliferation, survival, and membrane ruffling in C2C12 myoblasts. J. Cell Physiol. 187, 96–108.

Damato, B.E., Allan, D., Murray, S.B., Lee, W.R., 1984. Senile atrophy of the human lacrimal gland: the contribution of chronic inflammatory disease. Br. J. Ophthalmol. 68, 674–680.

Danielsen, A.G., Roth, R.A., 1996. Role of the juxtamembrane tyrosine in insulin receptor-mediated tyrosine phosphorylation of p60 endogenous substrates. Endocrinology 137, 5326–5331.

Davis, R.J., 1995. Transcriptional regulation by MAP kinases. Mol. Reprod. Dev. 42, 459–467.

Davis, S., Aldrich, T.H., Valenzuela, D.M., Wong, V.V., Furth, M.E., Squinto, S.P., Yancopoulos, G.D., 1991. The receptor for ciliary neurotrophic factor. Science 253, 59–63.

de, L.A., Fernandes, M.L., Saad, M.J., Velloso, L.A., 1999. Insulin induces tyrosine phosphorylation of the insulin receptor and SHC, and SHC/GRB2 association in cerebellum but not in forebrain cortex of rats. Brain Res. 826, 74–82.

De-Fraja, C., Conti, L., Govoni, S., Battaini, F., Cattaneo, E., 2000. STAT signalling in the mature and aging brain. Int. J. Dev. Neurosci. 18, 439–446.

Dogru, M., Katakami, C., Inoue, M., 2001. Tear function and ocular surface changes in noninsulin-dependent diabetes mellitus. Ophthalmology 108, 586–592.

Dominici, F.P., Hauck, S., Argentino, D.P., Bartke, A., Turyn, D., 2002. Increased insulin sensitivity and upregulation of insulin receptor, insulin receptor substrate (IRS)-1 and IRS-2 in liver of Ames dwarf mice. J. Endocrinol. 173, 81–94.

Dore, S., Kar, S., Rowe, W., Quirion, R., 1997. Distribution and levels of [125I]IGF-I, [125I]IGF-II and [125I]insulin receptor binding sites in the hippocampus of aged memory-unimpaired and -impaired rats. Neuroscience 80, 1033–1040.

Emanuelli, B., Peraldi, P., Filloux, C., Sawka-Verhelle, D., Hilton, D., Van Obberghen, E., 2000. SOCS-3 is an insulin-induced negative regulator of insulin signaling. J. Biol. Chem. 275, 15985–15991.

Ershler, W.B., 1993. Tumors and aging: the influence of age-associated immune changes upon tumor growth and spread. Adv. Exp. Med. Biol. 330, 77–92.

Espat, N.J., Auffenberg, T., Rosenberg, J.J., Rogy, M., Martin, D., Fang, C.H., Hasselgren, P.O., Copeland, E.M., Moldawer, L.L., 1996. Ciliary neurotrophic factor is catabolic and shares with IL-6 the capacity to induce an acute phase response. Am. J. Physiol. 271, R185–R190.

Evans, J.L., Goldfine, I.D., Maddux, B.A., Grodsky, G.M., 2002. Oxidative stress and stress-activated signaling pathways: a unifying hypothesis of type 2 diabetes. Endocr. Rev. 23, 599–622.

Fantuzzi, G., Benigni, F., Sironi, M., Conni, M., Carelli, M., Cantoni, L., Shapiro, L., Dinarello, C.A., Sipe, J.D., Ghezzi, P., 1995. Ciliary neurotrophic factor (CNTF) induces serum amyloid A, hypoglycaemia and anorexia, and potentiates IL-1 induced corticosterone and IL-6 production in mice. Cytokine 7, 150–156.

Fernandes, M.L., Saad, M.J., Velloso, L.A., 2001. Effects of age on elements of insulin-signaling pathway in central nervous system of rats. Endocrine 16, 227–234.

Fink, R.I., Kolterman, O.G., Griffin, J., Olefsky, J.M., 1983. Mechanisms of insulin resistance in aging. J. Clin. Invest. 71, 1523–1535.

Flurkey, K., Papaconstantinou, J., Miller, R.A., Harrison, D.E., 2001. Lifespan extension and delayed immune and collagen aging in mutant mice with defects in growth hormone production. Proc. Natl. Acad. Sci. USA 98, 6736–6741.

Folli, F., Bonfanti, L., Renard, E., Kahn, C.R., Merighi, A., 1994. Insulin receptor substrate-1 (IRS-1) distribution in the rat central nervous system. J. Neurosci. 14, 6412–6422.

Folli, F., Saad, M.J., Backer, J.M., Kahn, C.R., 1992. Insulin stimulation of phosphatidylinositol 3-kinase activity and association with insulin receptor substrate 1 in liver and muscle of the intact rat. J. Biol. Chem. 267, 22171–22177.

Foulstone, E.J., Tavare, J.M., Gunn-Moore, F.J., 1999. Sustained phosphorylation and activation of protein kinase B correlates with brain-derived neurotrophic factor and insulin stimulated survival of cerebellar granule cells. Neurosci. Lett. 264, 125–128.

Frank, H.J., Pardridge, W.M., Morris, W.L., Rosenfeld, R.G., Choi, T.B., 1986. Binding and internalization of insulin and insulin-like growth factors by isolated brain microvessels. Diabetes 35, 654–661.

Frolich, L., Blum-Degen, D., Bernstein, H.G., Engelsberger, S., Humrich, J., Laufer, S., Muschner, D., Thalheimer, A., Turk, A., Hoyer, S., Zochling, R., Boissl, K.W., Jellinger, K., Riederer, P., 1998. Brain insulin and insulin receptors in aging and sporadic Alzheimer's disease. J. Neural Transm. 105, 423–438.

Gary, D.S., Mattson, M.P., 2001. Integrin signaling via the PI3-kinase-Akt pathway increases neuronal resistance to glutamate-induced apoptosis. J. Neurochem. 76, 1485–1496.

Girault, J.A., Chamak, B., Bertuzzi, G., Tixier, H., Wang, J.K., Pang, D.T., Greengard, P., 1992. Protein phosphotyrosine in mouse brain: developmental changes and regulation by epidermal growth factor, type I insulin-like growth factor, and insulin. J. Neurochem. 58, 518–528.

Goodman, M.N., Dluz, S.M., McElaney, M.A., Belur, E., Ruderman, N.B., 1983. Glucose uptake and insulin sensitivity in rat muscle: changes during 3–96 weeks of age. Am. J. Physiol. 244, E93–E100.

Hanninen, T., Reinikainen, K.J., Helkala, E.L., Koivisto, K., Mykkanen, L., Laakso, M., Pyorala, K., Riekkinen, P.J., 1994. Subjective memory complaints and personality traits in normal elderly subjects. J. Am. Geriatr. Soc. 42, 1–4.

Havrankova, J., Roth, J., Brownstein, M., 1978a. Insulin receptors are widely distributed in the central nervous system of the rat. Nature 272, 827–829.

Havrankova, J., Roth, J., Brownstein, M.J., 1979. Concentrations of insulin and insulin receptors in the brain are independent of peripheral insulin levels. Studies of obese and streptozotocin-treated rodents. J. Clin. Invest. 64, 636–642.

Havrankova, J., Schmechel, D., Roth, J., Brownstein, M., 1978b. Identification of insulin in rat brain. Proc. Natl. Acad. Sci. USA 75, 5737–5741.

Henderson, J.T., Seniuk, N.A., Richardson, P.M., Gauldie, J., Roder, J.C., 1994. Systemic administration of ciliary neurotrophic factor induces cachexia in rodents. J. Clin. Invest. 93, 2632–2638.

Hilton, D.J., 1999. Negative regulators of cytokine signal transduction. Cell. Mol. Life Sci. 55, 1568–1577.

Holzenberger, M., Dupont, J., Ducos, B., Leneuve, P., Geloen, A., Even, P.C., Cervera, P., Le Bouc, Y., 2003. IGF-1 receptor regulates lifespan and resistance to oxidative stress in mice. Nature 421, 182–187.

Howard, J.K., Lord, G.M., Matarese, G., Vendetti, S., Ghatei, M.A., Ritter, M.A., Lechler, R.I., Bloom, S.R., 1999. Leptin protects mice from starvation-induced lymphoid atrophy and increases thymic cellularity in ob/ob mice. J. Clin. Invest. 104, 1051–1059.

Hoyer, S., 1991. Abnormalities of glucose metabolism in Alzheimer's disease. Ann. N.Y. Acad. Sci. 640, 53–58.

Hwang, J.B., Hernandez, J., Leduc, R., Frost, S.C., 2000. Alternative glycosylation of the insulin receptor prevents oligomerization and acquisition of insulin-dependent tyrosine kinase activity. Biochim. Biophys. Acta 1499, 74–84.

Inoue, M., Nakayama, C., Noguchi, H., 1996. Activating mechanism of CNTF and related cytokines. Mol. Neurobiol. 12, 195–209.

Kahn, C.R., 1994. Banting Lecture. Insulin action, diabetogenes, and the cause of type II diabetes. Diabetes 43, 1066–1084.

Kalant, N., Stewart, J., Kaplan, R., 1988. Effect of diet restriction on glucose metabolism and insulin responsiveness in aging rats. Mech. Ageing Dev. 46, 89–104.

Kanai, F., Ito, K., Todaka, M., Hayashi, H., Kamohara, S., Ishii, K., Okada, T., Hazeki, O., Ui, M., Ebina, Y., 1993. Insulin-stimulated GLUT4 translocation is relevant to the phosphorylation of IRS-1 and the activity of PI3-kinase. Biochem. Biophys. Res. Commun. 195, 762–768.

Kar, S., Chabot, J.G., Quirion, R., 1993. Quantitative autoradiographic localization of [125I]insulin-like growth factor I, [125I]insulin-like growth factor II, and [125I]insulin receptor binding sites in developing and adult rat brain. J. Comp. Neurol. 333, 375–397.

Kasuga, M., Karlsson, F.A., Kahn, C.R., 1982. Insulin stimulates the phosphorylation of the 95,000-dalton subunit of its own receptor. Science 215, 185–187.

Kelly, K.L., Ruderman, N.B., 1993. Insulin-stimulated phosphatidylinositol 3-kinase. Association with a 185-kDa tyrosine-phosphorylated protein (IRS-1) and localization in a low density membrane vesicle. J. Biol. Chem. 268, 4391–4398.

Kenyon, C., Chang, J., Gensch, E., Rudner, A., Tabtiang, R., 1993. A C. elegans mutant that lives twice as long as wild type. Nature 366, 461–464.

Kile, B.T., Nicola, N.A., Alexander, W.S., 2001. Negative regulators of cytokine signaling. Int. J. Hematol. 73, 292–298.

Klip, A., Paquet, M.R., 1990. Glucose transport and glucose transporters in muscle and their metabolic regulation. Diabetes Care 13, 228–243.

Kopelman, P.G., 2000. Obesity as a medical problem. Nature 404, 635–643.

Lambert, P.D., Anderson, K.D., Sleeman, M.W., Wong, V., Tan, J., Hijarunguru, A., Corcoran, T.L., Murray, J.D., Thabet, K.E., Yancopoulos, G.D., Wiegand, S.J., 2001. Ciliary neurotrophic factor activates leptin-like pathways and reduces body fat, without cachexia or rebound weight gain, even in leptin-resistant obesity. Proc. Natl. Acad. Sci. USA 98, 4652–4657.

Le Roith, D., Hendricks, S.A., Lesniak, M.A., Rishi, S., Becker, K.L., Havrankova, J., Rosenzweig, J.L., Brownstein, M.J., Roth, J., 1983. Insulin in brain and other extrapancreatic tissues of vertebrates and nonvertebrates. Adv. Metab. Disord. 10, 303–340.

Lee, J., Duan, W., Long, J.M., Ingram, D.K., Mattson, M.P., 2000. Dietary restriction increases the number of newly generated neural cells, and induces BDNF expression, in the dentate gyrus of rats. J. Mol. Neurosci. 15, 99–108.

Levy, D.E., Darnell, J.E., Jr., 2002. Stats: transcriptional control and biological impact. Nat. Rev. Mol. Cell. Biol. 3, 651–662.

Lin, K., Hsin, H., Libina, N., Kenyon, C., 2001. Regulation of the Caenorhabditis elegans longevity protein DAF-16 by insulin/IGF-1 and germline signaling. Nat. Genet. 28, 139–145.

Lindberg, K., Brown, M.E., Chaves, H.V., Kenyon, K.R., Rheinwald, J.G., 1993. In vitro propagation of human ocular surface epithelial cells for transplantation. Invest. Ophthalmol. Vis. Sci. 34, 2672–2679.

Lord, G.M., Matarese, G., Howard, J.K., Baker, R.J., Bloom, S.R., Lechler, R.I., 1998. Leptin modulates the T-cell immune response and reverses starvation-induced immunosuppression. Nature 394, 897–901.

Margolis, R.U., Altszuler, N., 1967. Insulin in the cerebrospinal fluid. Nature 215, 1375–1376.

Marks, J.L., Eastman, C.J., 1990. Ontogeny of insulin binding in different regions of the rat brain. Dev. Neurosci. 12, 349–358.

Marks, J.L., King, M.G., Baskin, D.G., 1991. Localization of insulin and type 1 IGF receptors in rat brain by in vitro autoradiography and in situ hybridization. Adv. Exp. Med. Biol. 293, 459–470.

Martin, D., Merkel, E., Tucker, K.K., McManaman, J.L., Albert, D., Relton, J., Russell, D.A., 1996. Cachectic effect of ciliary neurotrophic factor on innervated skeletal muscle. Am. J. Physiol. 271, R1422–R1428.

Martin, D., Salinas, M., Lopez-Valdaliso, R., Serrano, E., Recuero, M., Cuadrado, A., 2001. Effect of the Alzheimer amyloid fragment Abeta(25-35) on Akt/PKB kinase and survival of PC12 cells. J. Neurochem. 78, 1000–1008.

Matthaei, S., Benecke, H., Klein, H.H., Hamann, A., Kreymann, G., Greten, H., 1990. Potential mechanism of insulin resistance in ageing: impaired insulin-stimulated glucose transport due to a depletion of the intracellular pool of glucose transporters in Fischer rat adipocytes. J. Endocrinol. 126, 99–107.

Mattson, M.P., 1997. Cellular actions of beta-amyloid precursor protein and its soluble and fibrillogenic derivatives. Physiol. Rev. 77, 1081–1132.

Mattson, M.P., 2002. Brain evolution and lifespan regulation: conservation of signal transduction pathways that regulate energy metabolism. Mech. Ageing Dev. 123, 947–953.

Mattson, M.P., Duan, W., Maswood, N., 2002. How does the brain control lifespan? Ageing Res. Rev. 1, 155–165.

Mattson, M.P., Furukawa, K., 1997. Alzheimer's disease. Short precursor shortens memory. Nature 387, 457–458.

Mattson, M.P., Pedersen, W.A., Duan, W., Culmsee, C., Camandola, S., 1999. Cellular and molecular mechanisms underlying perturbed energy metabolism and neuronal degeneration in Alzheimer's and Parkinson's diseases. Ann. N.Y. Acad. Sci. 893, 154–175.

Mendez, R., Kollmorgen, G., White, M.F., Rhoads, R.E., 1997. Requirement of protein kinase C zeta for stimulation of protein synthesis by insulin. Mol. Cell. Biol. 17, 5184–5192.

Minichiello, L., Korte, M., Wolfer, D., Kuhn, R., Unsicker, K., Cestari, V., Rossi-Arnaud, C., Lipp, H.P., Bonhoeffer, T., Klein, R., 1999. Essential role for TrkB receptors in hippocampus-mediated learning. Neuron 24, 401–414.

Mytilineou, C., Werner, P., Molinari, S., Di Rocco, A., Cohen, G., Yahr, M.D., 1994. Impaired oxidative decarboxylation of pyruvate in fibroblasts from patients with Parkinson's disease. J. Neural Transm. Park. Dis. Dement. Sect. 8, 223–228.

Nadiv, O., Cohen, O., Zick, Y., 1992. Defects of insulin's signal transduction in old rat livers. Endocrinology 130, 1515–1524.

Nadiv, O., Shinitzky, M., Manu, H., Hecht, D., Roberts, C.T., Jr., LeRoith, D., Zick, Y., 1994. Elevated protein tyrosine phosphatase activity and increased membrane viscosity are associated with impaired activation of the insulin receptor kinase in old rats. Biochem. J. 298, 443–450.

Nakamura, T., Muraoka, S., Sanokawa, R., Mori, N., 1998. N-Shc and Sck, two neuronally expressed Shc adapter homologs. Their differential regional expression in the brain and roles in neurotrophin and Src signaling. J. Biol. Chem. 273, 6960–6967.

Nishimura, H., Kuzuya, H., Okamoto, M., Yoshimasa, Y., Yamada, K., Ida, T., Kakehi, T., Imura, H., 1988. Change of insulin action with aging in conscious rats determined by euglycemic clamp. Am. J. Physiol. 254, E92–E98.

Nussbaum, R.L., Ellis, C.E., 2003. Alzheimer's disease and Parkinson's disease. N. Engl. J. Med. 348, 1356–1364.

Ortiz, J., Harris, H.W., Guitart, X., Terwilliger, R.Z., Haycock, J.W., Nestler, E.J., 1995. Extracellular signal-regulated protein kinases (ERKs) and ERK kinase (MEK) in brain: regional distribution and regulation by chronic morphine. J. Neurosci. 15, 1285–1297.

Oxenkrug, G., Requintina, P., Bachurin, S., 2001. Antioxidant and antiaging activity of N-acetylserotonin and melatonin in the in vivo models Ann. N.Y. Acad. Sci. 939, 190–199.

Paez-Espinosa, E.V., Rocha, E.M., Velloso, L.A., Boschero, A.C., Saad, M.J., 1999. Insulin-induced tyrosine phosphorylation of Shc in liver, muscle and adipose tissue of insulin resistant rats. Mol. Cell. Endocrinol. 156, 121–129.

Pardridge, W.M., 1986. Receptor-mediated peptide transport through the blood–brain barrier. Endocr. Rev. 7, 314–330.

Parkes, T.L., Elia, A.J., Dickinson, D., Hilliker, A.J., Phillips, J.P., Boulianne, G.L., 1998. Extension of Drosophila lifespan by overexpression of human SOD1 in motorneurons. Nat. Genet. 19, 171–174.

Patti, M.E., Kahn, C.R., 1998. The insulin receptor – a critical link in glucose homeostasis and insulin action. J. Basic. Clin. Physiol. Pharmacol. 9, 89–109.

Pawelec, G., Hirokawa, K., Fulop, T., 2001. Altered T cell signalling in ageing. Mech. Ageing Dev. 122, 1613–1637.

Pecile, A., Muller, E., Falconi, G., Martini, L., 1965. Growth hormone-releasing activity of hypothalamic extracts at different ages. Endocrinology 77, 241–246.

Pedersen, W.A., Culmsee, C., Ziegler, D., Herman, J.P., Mattson, M.P., 1999. Aberrant stress response associated with severe hypoglycemia in a transgenic mouse model of Alzheimer's disease. J. Mol. Neurosci. 13, 159–165.

Pellegrini, G., Traverso, C.E., Franzi, A.T., Zingirian, M., Cancedda, R., De Luca, M., 1997. Long-term restoration of damaged corneal surfaces with autologous cultivated corneal epithelium. Lancet 349, 990–993.

Peralta, S., Carrascosa, J.M., Gallardo, N., Ros, M., Arribas, C., 2002. Ageing increases SOCS-3 expression in rat hypothalamus: effects of food restriction. Biochem. Biophys. Res. Commun. 296, 425–428.

Pessin, J.E., Saltiel, A.R., 2000. Signaling pathways in insulin action: molecular targets of insulin resistance. J. Clin. Invest. 106, 165–169.

Peyssonnaux, C., Eychene, A., 2001. The Raf/MEK/ERK pathway: new concepts of activation. Biol. Cell 93, 53–62.

Pomerance, M., Gavaret, J.M., Jacquemin, C., Matricon, C., Toru-Delbauffe, D., Pierre, M., 1988. Insulin and insulin-like growth factor 1 receptors during postnatal development of rat brain. Brain Res. 470, 77–83.

Portero-Otin, M., Pamplona, R., Bellmunt, M.J., Ruiz, M.C., Prat, J., Salvayre, R., Negre-Salvayre, A., 2002. Advanced glycation end product precursors impair epidermal growth factor receptor signaling. Diabetes 51, 1535–1542.

Potau, N., Escofet, M.A., Martinez, M.C., 1991. Ontogenesis of insulin receptors in human cerebral cortex. J. Endocrinol. Invest. 14, 53–58.

Rajala, R.V., McClellan, M.E., Ash, J.D., Anderson, R.E., 2002. In vivo regulation of phosphoinositide 3-kinase in retina through light-induced tyrosine phosphorylation of the insulin receptor beta-subunit. J. Biol. Chem. 277, 43319–43326.

Ribon, V., Herrera, R., Kay, B.K., Saltiel, A.R., 1998. A role for CAP, a novel, multifunctional Src homology 3 domain-containing protein in formation of actin stress fibers and focal adhesions. J. Biol. Chem. 273, 4073–4080.

Ribon, V., Saltiel, A.R., 1997. Insulin stimulates tyrosine phosphorylation of the proto-oncogene product of c-Cbl in 3T3-L1 adipocytes. Biochem. J. 324, 839–845.

Rocha, E.M., Carvalho, C.R., Saad, M.J., Velloso, L.A., 2002a. The influence of aging in the insulin-signaling system in rat exocrine glands. Adv. Exp. Med. Biol. 506, 27–31.

Rocha, E.M., Cunha, D.A., Carneiro, E.M., Boschero, A.C., Saad, M.J., Velloso, L.A., 2002b. Identification of insulin in the tear film and insulin receptor and IGF-1 receptor on the human ocular surface. Invest. Ophthalmol. Vis. Sci. 43, 963–967.

Rocha, E.M., Cunha, D.A., Carneiro, E.M., Boschero, A.C., Saad, M.J., Velloso, L.A., 2002c. Insulin, insulin receptor and insulin-like growth factor-I receptor on the human ocular surface. Adv. Exp. Med. Biol. 506, 607–610.

Rocha, E.M., de, M.L.M.H., Carvalho, C.R., Saad, M.J., Velloso, L.A., 2000. Characterization of the insulin-signaling pathway in lacrimal and salivary glands of rats. Curr. Eye Res. 21, 833–842.

Rossner, S., 2001. Obesity in the elderly – a future matter of concern? Obes. Rev. 2, 183–188.

Rother, K.I., Imai, Y., Caruso, M., Beguinot, F., Formisano, P., Accili, D., 1998. Evidence that IRS-2 phosphorylation is required for insulin action in hepatocytes. J. Biol. Chem. 273, 17491–17497.

Rotwein, P., Burgess, S.K., Milbrandt, J.D., Krause, J.E., 1988. Differential expression of insulin-like growth factor genes in rat central nervous system. Proc. Natl. Acad. Sci. USA 85, 265–269.

Ruderman, N.B., Kapeller, R., White, M.F., Cantley, L.C., 1990. Activation of phosphatidylinositol 3-kinase by insulin. Proc. Natl. Acad. Sci. USA 87, 1411–1415.

Rui, L., Yuan, M., Frantz, D., Shoelson, S., White, M.F., 2002. SOCS-1 and SOCS-3 block insulin signaling by ubiquitin-mediated degradation of IRS1 and IRS2. J. Biol. Chem. 277, 42394–42398.

Saad, M.J., 1994. Molecular mechanisms of insulin resistance. Braz. J. Med. Biol. Res. 27, 941–957.

Saad, M.J., Velloso, L.A., Carvalho, C.R., 1995. Angiotensin II induces tyrosine phosphorylation of insulin receptor substrate 1 and its association with phosphatidylinositol 3-kinase in rat heart. Biochem. J. 310, 741–744.

Sakaue, H., Ogawa, W., Takata, M., Kuroda, S., Kotani, K., Matsumoto, M., Sakaue, M., Nishio, S., Ueno, H., Kasuga, M., 1997. Phosphoinositide 3-kinase is required for insulin-induced but not for growth hormone- or hyperosmolarity-induced glucose uptake in 3T3-L1 adipocytes. Mol. Endocrinol. 11, 1552–1562.

Saltiel, A.R., Kahn, C.R., 2001. Insulin signalling and the regulation of glucose and lipid metabolism. Nature 414, 799–806.

Sasaoka, T., Kobayashi, M., 2000. The functional significance of Shc in insulin signaling as a substrate of the insulin receptor. Endocr. J. 47, 373–381.

Schein, O.D., Munoz, B., Tielsch, J.M., Bandeen-Roche, K., West, S., 1997. Prevalence of dry eye among the elderly. Am. J. Ophthalmol. 124, 723–728.

Schwartz, M.W., Figlewicz, D.P., Baskin, D.G., Woods, S.C., Porte, D., Jr., 1992. Insulin in the brain: a hormonal regulator of energy balance. Endocr. Rev. 13, 387–414.

Sell, D.R., Monnier, V.M., 1990. End-stage renal disease and diabetes catalyze the formation of a pentose-derived crosslink from aging human collagen. J. Clin. Invest. 85, 380–384.

Seo, J.H., Rah, J.C., Choi, S.H., Shin, J.K., Min, K., Kim, H.S., Park, C.H., Kim, S., Kim, E.M., Lee, S.H., Lee, S., Suh, S.W., Suh, Y.H., 2002. Alpha-synuclein regulates neuronal survival via Bcl-2 family expression and PI3/Akt kinase pathway. FASEB J. 16, 1826–1828.

Sims, N.A., Clement-Lacroix, P., Da Ponte, F., Bouali, Y., Binart, N., Moriggl, R., Goffin, V., Coschigano, K., Gaillard-Kelly, M., Kopchick, J., Baron, R., Kelly, P.A., 2000. Bone homeostasis in growth hormone receptor-null mice is restored by IGF-I but independent of Stat5. J. Clin. Invest. 106, 1095–1103.

Singh, M., Bachhawat, B.K., 1965. The distribution and variation with age of different uronic acid-containing mucopolysaccharides in brain. J. Neurochem. 12, 519–525.

Song, J., Wu, L., Chen, Z., Kohanski, R.A., Pick, L., 2003. Axons guided by insulin receptor in Drosophila visual system. Science 300, 502–505.

Sonntag, W.E., Lynch, C.D., Bennett, S.A., Khan, A.S., Thornton, P.L., Cooney, P.T., Ingram, R.L., McShane, T., Brunso-Bechtold, J.K., 1999. Alterations in insulin-like growth factor-1 gene and protein expression and type 1 insulin-like growth factor receptors in the brains of ageing rats. Neuroscience 88, 269–279.

Sreebny, L.M., Valdini, A., 1988. Xerostomia. Part I: Relationship to other oral symptoms and salivary gland hypofunction. Oral Surg. Oral Med. Oral Pathol. 66, 451–458.

Steinman, L., Conlon, P., Maki, R., Foster, A., 2003. The intricate interplay among body weight, stress, and the immune response to friend or foe. J. Clin. Invest. 111, 183–185.

Sullivan, D.A., Hann, L.E., Yee, L., Allansmith, M.R., 1990. Age- and gender-related influence on the lacrimal gland and tears. Acta Ophthalmol. (Copenh) 68, 188–194.

Sullivan, D.A., Wickham, L.A., Rocha, E.M., Kelleher, R.S., da Silveira, L.A., Toda, I., 1998. Influence of gender, sex steroid hormones, and the hypothalamic-pituitary axis on the structure and function of the lacrimal gland. Adv. Exp. Med. Biol. 438, 11–42.

Tartaglia, L.A., Dembski, M., Weng, X., Deng, N., Culpepper, J., Devos, R., Richards, G.J., Campfield, L.A., Clark, F.T., Deeds, J., et al., 1995. Identification and expression cloning of a leptin receptor, OB-R. Cell 83, 1263–1271.

Tatar, M., Bartke, A., Antebi, A., 2003. The endocrine regulation of aging by insulin-like signals. Science 299, 1346–1351.

Tatar, M., Khazaeli, A.A., Curtsinger, J.W., 1997. Chaperoning extended life. Nature 390, 30.

Tatar, M., Kopelman, A., Epstein, D., Tu, M.P., Yin, C.M., Garofalo, R.S., 2001. A mutant Drosophila insulin receptor homolog that extends life-span and impairs neuroendocrine function. Science 292, 107–110.

Tonra, J.R., Ono, M., Liu, X., Garcia, K., Jackson, C., Yancopoulos, G.D., Wiegand, S.J., Wong, V., 1999. Brain-derived neurotrophic factor improves blood glucose control and alleviates fasting hyperglycemia in C57BLKS-Lepr(db)/lepr(db) mice. Diabetes 48, 588–594.

Torsoni, M.A., Carvalheira, J.B., Pereira-Da-Silva, M., De Carvalho-Filho, M.A., Saad, M.J., Velloso, L.A., 2003. Molecular and functional resistance to insulin in hypothalamus of rats exposed to cold. Am. J. Physiol. Endocrinol. Metab. 18, 18.

Tower, J., 2000. Transgenic methods for increasing Drosophila life span. Mech. Ageing Dev. 118, 1–14.

Velloso, L.A., Carvalho, C.R., Rojas, F.A., Folli, F., Saad, M.J., 1998. Insulin signalling in heart involves insulin receptor substrates-1 and-2, activation of phosphatidylinositol 3-kinase and the JAK 2-growth related pathway. Cardiovasc. Res. 40, 96–102.

Velloso, L.A., Folli, F., Sun, X.J., White, M.F., Saad, M.J., Kahn, C.R., 1996. Cross-talk between the insulin and angiotensin signaling systems. Proc. Natl. Acad. Sci. USA 93, 12490–12495.

Vlassara, H., Bucala, R., Striker, L., 1994. Pathogenic effects of advanced glycosylation: biochemical, biologic, and clinical implications for diabetes and aging. Lab. Invest. 70, 138–151.

Vlassara, H., Fuh, H., Makita, Z., Krungkrai, S., Cerami, A., Bucala, R., 1992. Exogenous advanced glycosylation end products induce complex vascular dysfunction in normal animals: a model for diabetic and aging complications. Proc. Natl. Acad. Sci. USA 89, 12043–12047.

Wada, T., Sasaoka, T., Ishiki, M., Hori, H., Haruta, T., Ishihara, H., Kobayashi, M., 1999. Role of the Src homology 2 (SH2) domain and C-terminus tyrosine phosphorylation sites of SH2-containing inositol phosphatase (SHIP) in the regulation of insulin-induced mitogenesis. Endocrinology 140, 4585–4594.

Weindruch, R., Sohal, R.S., 1997. Seminars in medicine of the Beth Israel Deaconess Medical Center. Caloric intake and aging. N. Engl. J. Med. 337, 986–994.

White, M.F., 2002. IRS proteins and the common path to diabetes. Am. J. Physiol. Endocrinol. Metab. 283, E413–E422.

White, M.F., Kahn, C.R., 1994. The insulin signaling system. J. Biol. Chem. 269, 1–4.

Woods, S.C., Lotter, E.C., McKay, L.D., Porte, D., Jr., 1979. Chronic intracerebroventricular infusion of insulin reduces food intake and body weight of baboons. Nature 282, 503–505.

Yamaguchi, A., Tamatani, M., Matsuzaki, H., Namikawa, K., Kiyama, H., Vitek, M.P., Mitsuda, N., Tohyama, M., 2001. Akt activation protects hippocampal neurons from apoptosis by inhibiting transcriptional activity of p53. J. Biol. Chem. 276, 5256–5264.

Yenush, L., Zanella, C., Uchida, T., Bernal, D., White, M.F., 1998. The pleckstrin homology and phosphotyrosine binding domains of insulin receptor substrate 1 mediate inhibition of apoptosis by insulin. Mol. Cell. Biol. 18, 6784–6794.

Zahniser, N.R., Goens, M.B., Hanaway, P.J., Vinych, J.V., 1984. Characterization and regulation of insulin receptors in rat brain. J. Neurochem. 42, 1354–1362.

Zaia, A., Piantanelli, L., 2000. Insulin receptors in the brain cortex of aging mice. Mech. Ageing Dev. 113, 227–232.

Zatechka, S.D., Jr., Lou, M.F., 2002. Studies of the mitogen-activated protein kinases and phosphatidylinositol-3 kinase in the lens. 1. The mitogenic and stress responses. Exp. Eye Res. 74, 703–717.

Zecchin, H.G., Bezerra, R.M., Carvalheira, J.B., Carvalho-Filho, M.A., Metze, K., Franchini, K.G., Saad, M.J., 2003. Insulin signalling pathways in aorta and muscle from two animal models of insulin resistance – the obese middle-aged and the spontaneously hypertensive rats. Diabetologia 5, 5.

Zhao, W.Q., Alkon, D.L., 2001. Role of insulin and insulin receptor in learning and memory. Mol. Cell. Endocrinol. 177, 125–134.

Zhao, W., Chen, H., Xu, H., Moore, E., Meiri, N., Quon, M.J., Alkon, D.L., 1999. Brain insulin receptors and spatial memory. Correlated changes in gene expression, tyrosine phosphorylation, and signaling molecules in the hippocampus of water maze trained rats. J. Biol. Chem. 274, 34893–34902.

Zhao, W.Q., Ravindranath, L., Mohamed, A.S., Zohar, O., Chen, G.H., Lyketsos, C.G., Etcheberrigaray, R., Alkon, D.L., 2002. MAP kinase signaling cascade dysfunction specific to Alzheimer's disease in fibroblasts. Neurobiol. Dis. 11, 166–183.

**Advances in
Cell Aging and
Gerontology**

Age-related changes in synaptic phosphorylation and dephosphorylation

Thomas C. Foster

*McKnight Chair for Brain Research in Memory Loss, Department of Neuroscience,
Evelyn F. & William L. McKnight Brain Institute, University of Florida College of Medicine,
PO Box 100244, Gainesville, FL 32610-0244.
Tel.: (352) 392-4359; fax: (352) 392-8347.
E-mail address: foster@mbi.ufl.edu*

Contents

1. Introduction

The second messenger concept of cell signaling involving protein phosphorylation was proposed almost 50 years ago (Fisher and Kreb, 1955). Posttranslational modification of proteins through phosphorylation provides a rapid and relatively long-term mechanism for modifying physiological function. The rapid switching of protein function is observed for nearly all cellular processes including regulation of neuronal properties that are important for information processing. Indeed, the pervasive nature of protein phosphorylation is particularly evident in the control of synaptic transmission properties and the induction of synaptic plasticity (Tokuda and Hatase, 1998; Foster, 1999). An array of serine/threonine and tyrosine kinases and phosphatases are expressed in the brain and many are enriched at synaptic sites. The balance of kinase and phosphatase activity is under the control of several signaling cascades to influence synaptic transmission through altered transmitter release, postsynaptic receptor responsiveness, and

Advances in Cell Aging and Gerontology, vol. 16, 133–152
DOI: 10.1016/S1566-3124(04)16006-9

synaptic structure. Furthermore, the phosphorylation state of key proteins involved in gene regulation, such as transcription factors, nuclear receptors, and chromatin structures, influence synaptic turnover and maintenance.

Much of the research on age-related changes in protein phosphorylation at the synapse has focused on the hippocampus, due to the rich history of literature on synaptic plasticity in this structure and theoretical considerations of changes in hippocampal synaptic transmission in mediating cognitive decline with advanced age (Foster, 1999). In fact, decreased hippocampal synaptic transmission is well-established marker of brain aging, which correlates with decreased memory function (Barnes et al., 2000). Evidence for protein phosphorylation mechanisms in mediating the reduction of synaptic strength with age was provided by research demonstrating that the broad-spectrum serine/threonine kinase inhibitor, H7, differentially reduced CA3–CA1 synaptic strength in young adult, relative to aged rats (Norris et al., 1998). In the same study, extracellular application of the protein phosphatase 1 (PP1) and protein phosphatase 2A (PP2A) inhibitor, calyculin A, increased synaptic strength in aged animals relative to young adults. While the results point to a shift in the balance of kinases and phosphatases in mediating the age-related decrease in synaptic transmission, the report left open the question of which phosphoproteins might be involved.

2. Phosphorylation state and synaptic strength

2.1. Presynaptic mechanisms

Altered synaptic transmission due to a shift in enzyme activity is likely to involve several mechanisms, both presynaptic and postsynaptic. For example, serine/threonine kinase phosphorylation of tyrosine hydroxylase increases catecholamine synthesis and phosphorylation of this rate-limiting enzyme is decreased in aged animals suggesting reduced production of transmitter (Unnerstall and Ladner, 1994). In general, transmitter release depends on the balance of kinase/phosphatase activity such that activation of kinases is associated with increased transmitter release (Fig. 1A). Thus, the release of transmitter is increased by activation of protein kinase C (PKC) (Dekker et al., 1991; Robinson, 1991) and Ca^{2+}-dependent release is likely modified by the activity of Ca^{2+}/calmodulin kinase II (CaMKII) (He et al., 2000; Verona et al., 2000), and adenylyl cyclase-dependent protein kinase A (PKA) (Evans et al., 2001; Hilfiker et al., 2001). Moreover, most studies using phosphatase inhibitors suggest that dephosphorylation suppresses transmitter release (Sim et al., 1993; Vickroy et al., 1995; Yakel, 1997; Lin and Lin-Shiau, 1999). The increase in transmitter release associated with kinase activation may involve a shift in the pool of releasable and reserve vesicles through the phosphorylation of synapsin proteins, which regulate the interaction of vesicles with cytoskeletal proteins (Hilfiker et al., 1999). Furthermore, kinase activity, acting on vesicle proteins, may alter synaptic vesicle trafficking and the turnover due to the altered interactions of vesicle proteins with their transport effector proteins (Lin and Scheller, 2000). Moreover, phosphorylation of soluble NSF attachment protein

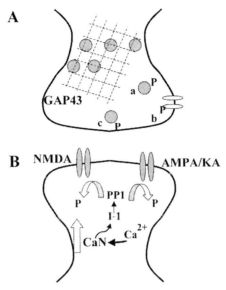

Fig. 1. (A) Activation of serine/threonine kinases is associated with an increased release of transmitter. Several mechanisms are thought to mediate the kinase-induced increase in transmitter release. (a) A reserve pool of vesicles (gray circles) is maintained through the interaction of the vesicle with cytoskeletal proteins. During increase of neuronal activity, kinases phosphorylate synapsins resulting in a transfer of the vesicle to a releasable pool. (b) Phosphorylation of Ca^{2+} or K^+ channels changes the flux of Ca^{2+} into the axon terminal. (c) Vesicle fusion and the responsiveness of release to Ca^{2+} may be sensitive to the phosphorylation of SNARE proteins. Finally, phosphorylation of GAP43 is involved in cytoskeletal remodeling and neurite outgrowth. An age-related impairment in kinase activation would limit the functional plasticity of the presynaptic elements. (B) Changes in Ca^{2+} homeostatic mechanisms result in a modest rise in intracellular Ca^{2+} leading to increased activity of CaN. In turn, CaN increases the activity of PP1 through dephosphorylation of inhibitor 1 (I-1). Increased activity of both serine/threonine phosphatases can decrease synaptic transmission and alter synaptic plasticity through dephosphorylation of AMPA/KA and NMDA receptors, respectively.

receptor (SNARE) proteins controls the responsiveness of release to Ca^{2+} (Verona et al., 2000) as well as vesicle fusion (Lin and Scheller, 2000), providing potential mechanisms for direct regulation of exocytosis by phosphorylation. Finally, transmitter release is dependent on the waveform characteristics of the action potential such that presynaptic channel proteins act to regulate transmitter release properties through the phosphorylation of specific Ca^{2+} or K^+ channels (Bartschat and Rhodes, 1995; Hell et al., 1995; Roeper et al., 1997; Tomizawa et al., 2002). Indeed, decreased cell excitability in aged animals is associated with increased Ca^{2+} channel function possibly due to increased phosphatase activity (Norris et al., 2002).

Very little is known concerning changes in the phosphorylation state of specific presynaptic proteins during aging. Furthermore, for studies that have focused on specific phosphoproteins, there is no consensus as to how basal phosphorylation state changes with age (Barnes et al., 1988; Gianotti et al., 1993; Battaini et al., 1995). Moreover, evidence from several sources suggests that decreased basal synaptic transmission is not due to a change in transmitter-release properties

(Segovia et al., 2001). In contrast, aged animals may exhibit deficits in phosphorylation processes that rapidly modify transmitter release, hence influencing synaptic plasticity. For instance, an age-related impairment is observed in stimulation-induced phosphorylation of GAP-43 (Gianotti et al., 1993) and synapsin (Parfitt et al., 1991; Eckles et al., 1997), presynaptic proteins involved in regulating transmitter release and the formation and maintenance of synaptic contacts (Ramakers et al., 1997; Ferreira and Rapoport, 2002). A reduction in the ability to activate presynaptic phosphorylation mechanisms would limit the functional plasticity of the system and may influence synaptic turnover, promoting the loss of synaptic contacts.

2.2. AMPA/kainate receptors

In addition to a loss of synapses, surviving synaptic contacts appear to exhibit a decrease in postsynaptic responsiveness to glutamate during aging, suggesting either a reduction in the number, composition, or functional state of the receptors (Foster, 2002). Glutamate is the predominate excitatory neurotransmitter in the hippocampus, and the level of postsynaptic depolarization by glutamate is a function of the phosphorylation state of ionotropic receptors, particularly the α-amino-3-hydroxy-5-methyl-4-isoxalone propionate (AMPA) and kainate receptors. For example, phosphorylation of AMPA receptors rapidly increases postsynaptic responsiveness to transmitter (Greengard et al., 1991; Blackstone et al., 1994; Roche et al., 1996; Barria et al., 1997; Mammen et al., 1997; Kameyama et al., 1998; Lee et al., 1998; Tokuda and Hatase, 1998; Derkach et al., 1999; Banke et al., 2000; Lee et al., 2000).

AMPA receptors are heteromeric complexes, which in the hippocampus consist mainly of GluR1–GluR2 or GluR2–GluR3 tetramers (Song and Huganir, 2002). Phosphorylation sites on the GluR1 subunit have been identified at two serine residues of the C-terminus, serine 845 (Ser845) and serine 831 (Ser831). Interestingly, the phosphorylation state of these two sites appears to be controlled by different kinases and phosphatases permitting differential enzyme activity to regulate the expression of various forms of use-dependent synaptic plasticity. For example, phosphorylation of AMPA receptors contributes to the enhancement of synaptic transmission following induction of long-term potentiation (LTP) (Roche et al., 1996; Barria et al., 1997; Mammen et al., 1997; Derkach et al., 1999; Lee et al., 2000; Huang et al., 2001) and dephosphorylation is associated with decreased transmission following induction of long-term depression (LTD) (Roche et al., 1996; Mammen et al., 1997; Kameyama et al., 1998; Lee et al., 1998).

Increasing the activity of PKA in hippocampal cell cultures and acutely dissociated CA1 pyramidal neurons from neonates results in increased glutamate responsiveness. PKA activation increases the response of kainate receptors through phosphorylation of Ser684 of the GluR6 subunit (Raymond et al., 1993). Similarly, PKA-mediated phosphorylation of GluR1 Ser845 enhances currents through AMPA receptors resulting in an increase in the amplitude for spontaneous miniature synaptic events (Greengard et al., 1991; Wang et al., 1991). Moreover,

PKA activity and Ser845 phosphorylation may be involved in enhanced synaptic transmission due to neuromodulators (Gu and Moss, 1996; Svenningsson et al., 2002). However, other researchers have suggested that Ser845 of the AMPA receptor is phosphorylated under basal conditions, at least in younger animals (Roche et al., 1996; Kameyama et al., 1998; Lee et al., 1998). Rather, the dephosphorylation and phosphorylation of Ser845 is thought to mediate expression of LTD and reversal of LTD, respectively (Roche et al., 1996; Mammen et al., 1997; Kameyama et al., 1998; Lee et al., 1998, 2000).

The induction of LTD and subsequent dephosphorylation of Ser845 on the GluR1 subunit is mediated by increased activity of protein phosphatases, calcineurin (CaN or PP2B), and PP1 (Mulkey et al., 1994; Hodgkiss and Kelly, 1995; Torii et al., 1995; Norris et al., 1998; Ehlers, 2000; Lin et al., 2000; Ramakers et al., 2000). During pattern stimulation to induce LTD, a modest rise in Ca^{2+} induces CaN release from membrane-binding proteins into the cytosol where it is active (Graef et al., 1999; Ghetti and Heinemann, 2000; Foster et al., 2001). CaN dephosphorylates several classes of proteins including GluR1 sites on the AMPA receptor (Banke et al., 2000). Moreover, CaN dephosphorylation of inhibitor 1 increases PP1 activity resulting in a phosphatase cascade. Furthermore, these phosphatases can dephosphorylate some kinases, reducing their activity. Thus, a small increase in intracellular Ca^{2+} can initiate a cascade that shifts the balance of kinase and phosphatase activity, favoring phosphatase activity.

In contrast, a large rise in intracellular Ca^{2+} can shift the balance of kinase and phosphatase activity to favor kinase activation, particularly PKC and CaMKII. Similar to PKA, activation of PKC and CaMKII is associated with an increase in the current through the AMPA receptor (Roche et al., 1996; Derkach et al., 1999). However, the activation of either of these kinases results in phosphorylation of the Ser831 residue of the GluR1 subunit (Roche et al., 1996; Mammen et al., 1997). Indeed, induction of LTP involves a substantial rise in intracellular Ca^{2+}, the activation of CaMKII and PKC, and phosphorylation of Ser831 (Barria et al., 1997; Lee et al., 2000), and reversal of LTP due to patterned neural activity is linked with dephosphorylation of Ser831 through the CaN-PP1 cascade (Lee et al., 2000). Thus, synaptic strength is determined by the history of neural activity, which mediates the activity of serine/threonine protein kinases and phosphatases, to regulate the phosphorylation state of PKA and PKC/CaMKII sensitive sites on GluR1. In this way, changes in phosphorylation influence the postsynaptic responsiveness to glutamate to regulate the strength of synaptic transmission (Fig. 1B).

Figure 2 illustrates the results from studies using membrane-permeable phosphatase and kinase inhibitors, which suggest that the balance of postsynaptic kinase/phosphatase activity plays a role in maintaining basal level of synaptic strength and enzyme activity is shifted in favor of phosphatases during aging (Norris et al., 1998). Figure 3 provides some of the evidence for postsynaptic phosphorylation mechanisms in mediating the reduction of synaptic strength with age. Intracellular perfusion of CA1 pyramidal cells with the membrane-impermeable phosphatase inhibitor, microcystin, preferentially increased synaptic strength in

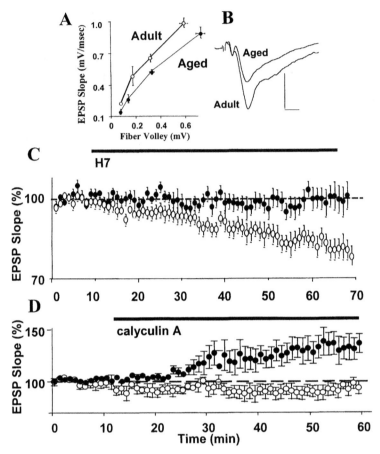

Fig. 2. (A) An age-related decrease in synaptic strength can be observed by plotting excitatory postsynaptic potential (EPSP) slope (mV/ms) against fiber volley amplitude (mV) for aged ($n = 13$, filled circles) and young adult ($n = 15$, open circles) animals. (B) Overlay of means from five consecutive extracellular waveforms recorded from both a single aged and a single young adult that exhibited similar fiber volley amplitudes but markedly reduced EPSP slopes, indicating a decrease in synaptic strength for aged slices. Calibration bars, 1 mV/5 ms. (C) The kinase inhibitor, H-7, depresses synaptic strength in slices from young adult, but not aged rats. Average data from aged ($n = 7$, filled circles) and adult ($n = 9$, open circles) slices that received H-7. (D) In contrast, the phosphatase inhibitor calyculin A increased synaptic transmission in aged ($n = 6$, filled circles), but not young adult ($n = 6$, open circles) slices. Figures are reprinted from Norris et al. (1998).

neurons from aged animals (Norris et al., 1998). Although, together, this work established the importance of the balance of kinase and phosphatase activity of the postsynaptic neurons in regulating basal synaptic strength during aging, it is unclear which specific kinases and phosphatases may be involved, what phosphoproteins mediate the age-related decline in synaptic strength, and whether presynaptic kinase/phosphatase activity is also altered.

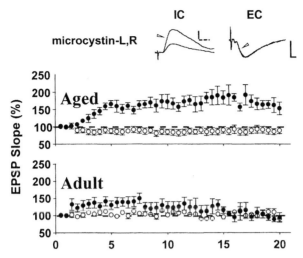

Fig. 3. Intracellular injection of microcystin enhances synaptic strength in an age-dependent fashion. Average data collected intracellularly (filled circles) and extracellularly (open circles) from 12 aged (*top trace*) and 12 adult (*bottom trace*) rat slices during intracellular recording with pipettes that contained 5 μM microcystin-L,R. *Inset*: Overlay of means from 10 consecutive responses collected intracellularly (IC) and extracellularly (EC) during the baseline and 40 min after (arrow head) impalement. Calibration bars, IC, 5 mV/10 ms; EC, 0.5 mV/5 ms. Figures are reprinted from Norris et al. (1998).

2.3. NMDA and metabotropic glutamate receptors

Age-related differences are observed in the propensity for hippocampal synapses to undergo modifications in transmission. This is particularly true for stimulation patterns that are near the threshold for induction of synaptic plasticity, such that LTP induction requires higher-frequency stimulation and LTD induction can be observed for lower-frequency stimulation. As such, the frequency-response function for older animals exhibits a pronounced plateau region in which LTD is observed at low frequencies that normally would not influence the response in young animals and no change in synaptic strength occurs for stimulation patterns that are near the threshold for induction of LTP (Foster, 1999). Thus, aging can be characterized by a shift in the balance of synaptic plasticity, with insufficient LTP induction and maintenance and excessive synaptic depression. This imbalance is thought to contribute to decreased synaptic strength at CA1 synapses (Foster and Norris, 1997) and the reduced transmission is maintained by the shift in the activity of the enzymes that underlie LTP and LTD (Foster, 2002; Foster and Kumar, 2002).

The influx of Ca^{2+} through the *N*-methyl-D-aspartate (NMDA) subtype glutamate receptor plays a central role in the regulation of induction of synaptic plasticity through the activation of Ca^{2+}-dependent kinases or phosphatases. Like the AMPA receptor, the function of the NMDA receptor is influenced by phosphorylation mechanisms. Tyrosine kinase activity increases NMDA receptor current (Heidinger et al., 2002), and receptor function is reduced by tyrosine phosphatase (Wang et al., 1996). Furthermore, the current through NMDA receptors is enhanced by PKC (Ben-Ari et al., 1992; Chen and Huang, 1992) and PKA activation (Raman et al., 1996), and inhibited by serine/threonine phosphatases including CaN and PP1 (Lieberman and Mody, 1994; Wang et al., 1994; Raman et al., 1996; Boxer et al., 1999). As such, the receptor-mediated rise in intracellular Ca^{2+} can activate Ca^{2+}-dependent signaling cascades that feedback to influence NMDA receptor function and the susceptibility to ensuing NMDA receptor-induced plasticity (Dingledine et al., 1999).

NMDA receptors are heteromers composed of NR1 and NR2 subunits. The NR1 subunit is central to proper receptor function and the NR1 gene can exhibit alternative splicing to produce eight receptor variants (Carroll and Zukin, 2002). Phosphorylation of serine residues within the alternatively spliced cassettes of the C-terminal tail (C1) of NR1 may influence receptor levels at the synapse. For example, phosphorylation of Ser896/897 of the NR1 subunit by PKC promotes receptor trafficking from the endoplasmic reticulum and insertion into the postsynaptic membrane over the course of several hours (Lan et al., 2001; Scott et al., 2001; Carroll and Zukin, 2002). In contrast, PKC phosphorylation of NR1 Ser890 within the C1 cassette may disrupt receptor aggregation altering the distribution of the receptor (Ehlers et al., 1995; Tingley et al., 1997) resulting in movement of receptors to extrasynaptic locations (Roche et al., 2001), and a reduction in receptor responsiveness (Markram and Segal, 1992).

NMDA receptor function can be influenced by phosphorylation of the NR2 subunit. Four distinct NR2 subunits (A–D) interact with the NR1 subunit and provide much of the modulatory control of receptor function (Dingledine et al., 1999). Evidence has been provided for direct NR2 phosphorylation of serine/ threonine (Liao et al., 2001) and tyrosine residues (Heidinger et al., 2002). However, the kinase-mediated enhancement is common for several NR1–NR2 subunit combinations in which specific phosphorylation sites are substituted (Swope et al., 1999). The results suggest that phosphorylation of proteins interacting with the NMDA receptor may mediate the changes in receptor function. For example, more rapid delivery of NMDA receptors may result from PKC interacting with trafficking proteins (Lan et al., 2001). Finally, NMDA responses are reduced by receptor internalization involving tyrosine dephosphorylation of NR2 units and this form of regulation may change over development/maturation with a shift in the level of specific NR2 subunits (Vissel et al., 2001; Li et al., 2002).

Metabotropic glutamate receptor (mGluR) activity can initiate intracellular signaling that results in a shift in the balance of protein kinase and phosphatase activity and is therefore involved in regulating a number of physiological processes related to synaptic transmission (Anwyl, 1999). The presynaptic group II and III mGluRs inhibit transmitter release and PKA activity impairs presynaptic mGluR function, increasing transmitter release, through phosphorylation of a Ser843 residue that is conserved across several mGluR subtypes (Schaffhauser et al., 2000; Cai et al., 2001). In contrast, group I mGluRs activate several postsynaptic second messenger systems including activation of PKC and tyrosine kinases to increase transmission and cell excitability. Phosphorylation state controls the function of type I receptors such that a decrease or increase in receptor function results from PKC and CaN activation, respectively (Alaluf et al., 1995; Minakami et al., 1997; Gereau and Heinemann, 1998; Alagarsamy et al., 1999; Ciruela et al., 1999; Dale et al., 2000). Furthermore, recent evidence indicates tyrosine phosphorylation of mGluR5, although the effect on receptor function is unclear (Orlando et al., 2002). Therefore, it is likely that mGluR function undergoes feedback regulation through serine/threonine and tyrosine phosphorylation.

Again, research on NMDA or mGluR phosphorylation during aging is rather sparse. There may be a reduction in NMDA receptors during aging, which would likely influence synaptic plasticity (Foster, 2002). A decrease in NMDA receptor phosphorylation has been noted in patients with Alzheimer's disease (Szc et al., 2001). In particular, a decline in tyrosine phosphorylation of NR2 subunits was correlated with memory impairments. Likewise, the effect of age on mGluR function and phosphorylation state is not clear. Differences in type I mGluR function may depend on the signaling cascade examined including phosphoinositide turnover or intracellular Ca^{2+} and subsequent PKC activation (Parent et al., 1995; Rahman et al., 1995; Pintor et al., 1998; Nicolle et al., 1999; Pintor et al., 2000; Attucci et al., 2002), or processes related to tyrosine kinase activation (Abdul-Ghani et al., 1996; Jouvenceau et al., 1997). There is some indication that expression of group II and III mGluRs may decrease in some brain regions (Magnusson, 1998; Simonyi et al., 2000). However, the effect on presynaptic function could be offset by a decrease in the activation of kinases in aged animals.

3. Age-related shift in kinase and phosphatase activity

3.1. Decreased kinase activity

While little is known concerning the phosphorylation state of synaptic proteins during aging, what is available would suggest dephosphorylation or the inability to stimulate the activity of kinases for both presynaptic (Parfitt et al., 1991; Gianotti et al., 1993; Battaini et al., 1995; Eckles et al., 1997) and postsynaptic neurons (Norris et al., 1998; Foster et al., 2001). In contrast, several groups have examined kinase and phosphatase activity and most indicate that the aging brain exhibits a decrease in kinase activity. However, conspicuous in this research is the considerable variability across animal models. In order to identify potential mechanisms for age-related changes in kinase/phosphatase activity, it is important to determine whether changes are due to altered expression, localization, or activation of enzymes. Much of the research has focused on Ca^{2+}-dependent enzymes due, in part, to the hypothesized dysregulation of Ca^{2+} homeostasis with advanced age.

CaMKII is a serine/threonine protein kinase that depends on Ca^{2+}/CaM for the initiation of activity and can undergo autophosphorylation, which influences the duration and targeting of kinase activity (Hudmon and Schulman, 2002). Thus, CaMKII phosphorylation can act as a marker of enzyme activation. In mouse models of aging, the expression of hippocampal CaMKII and the basal level of CaMKII phosphorylation are not altered (Armbrecht et al., 1999; Watson et al., 2002). However, the activity-dependent regulation of CaMKII transcription is depressed in aged rats (Davis et al., 2000). Furthermore, studies focused on aging rats indicate a decrease in the activation/translocation of CaMKII (Parfitt et al., 1991; Mullany et al., 1996; Eckles et al., 1997). The exact mechanism for the decrease in CaMKII activation remains to be elucidated; however, possible mechanisms include a change in CaMKII cofactors including oxidation of CaM (Gao et al., 1998; Mons et al., 2001) or alterations in the upstream-signaling cascade involving other phosphatases and kinases (Hudmon and Schulman, 2002).

A number of studies have found that basal brain activity of the serine–threonine kinase, PKC, is reduced and activation of hippocampal PKC is impaired, particularly in aged animals that exhibit memory deficits (Pascale et al., 1998a). There are several variants of PKC that represent a multigene family with at least 12 different isoforms grouped into three subclasses according to their regulatory properties. The first group is activated by Ca^{2+} and diacylglycerol (DAG) or phorbol ester and includes $PKC\alpha$, βI, βII, and γ. The second group (δ, ε, θ, μ, and η) is also activated by DAG and phorbol ester, but is Ca^{2+}-independent. Finally, ζ, ι, and λ are not responsive to Ca^{2+}, DAG, or phorbol ester.

Overall expression of various PKCs has been reported to remain constant (Mei et al., 1999) or to decrease (Mizutani et al., 1998) depending on the brain region examined. Others have reported that the localization of Ca^{2+}-dependent PKC is shifted with age such that no change in expression was noted for the hippocampus, cortex, and striatum of middle-aged rats; however, PKC activity shifted from the membrane/particulate and cytosolic fractions to increase in the nucleus (Battaini et al., 1995; La Porta and Comolli, 1999). Similarly, a loss of $PKC\gamma$ has been observed in the membrane fraction of the hippocampus in SAMP8 mice (Armbrecht et al., 1999).

Differences in the location of PKC are suggestive that the activation of PKC is altered, since translocation of PKC from the cytosol to the membrane is an integral step in kinase activation. Translocation of PKC due to neural activity is believed to play an important role in the initiation of kinase-dependent processes such as the induction of synaptic plasticity (Staak et al., 1995; Hrabetova and Sacktor, 2001; Lan et al., 2001) and learning (Van der Zee et al., 1997; Douma et al., 1998). A number of studies have reported that translocation of PKC from the cytosol to the membrane is reduced in the cortex and hippocampus of aged animals, suggesting a change in the regulatory mechanisms with age (Friedman and Wang, 1989; Meyer and Judkins, 1993; Meyer et al., 1994; Battaini et al., 1995; Undie et al., 1995). This apparent deficit in the activation of PKC has been correlated with the extent of memory impairments (Fordyce and Wehner, 1993; Colombo et al., 1997; Pascale et al., 1998b). The deficit in PKC activation may relate to changes in receptors signaling interactions such as the loss of phospholipase C activity (Mizutani et al., 1998). However, activating impairments are also observed following phorbol ester treatment suggesting that the deficits are not due to receptor activation and upstream-signaling mechanisms. A loss of the PKC-anchoring proteins, including the receptor for activated C kinase 1 (RACK-1), has been postulated to disrupt the PKC–membrane interaction during aging (Pascale et al., 1998a; Racchi et al., 2001).

Another major kinase of interest for synaptic function is PKA, whose activity is indirectly influenced by Ca^{2+}-regulation of adenylyl cyclase (AC). PKA exists as a tetrameric holoenzyme consisting of two catalytic subunits and two regulatory subunits (Brandon et al., 1997). The PKA regulatory subunits, type I or type II (RI and RII), are bound to A-kinase anchoring proteins (AKAPs), which in turn can be localized to synaptic cites by interactions with synaptic scaffolding proteins (Colledge et al., 2000). In addition, AKAPs bind other kinases and phosphatases (PKC and CaN) that influence PKA activity (Coghlan et al., 1995). PKA is released from the regulatory units to become activated in response to cellular events that stimulate cAMP production, such as a rise in intracellular Ca^{2+} (AC1 and 8) or through phosphorylation following activation of PKC (AC2 and 7). Interestingly, the level of Ca^{2+} that initiates AC1 activity inhibits AC9 activation due to Ca^{2+}-dependent activation of CaN (Antoni et al., 1998). Thus, the activation and specificity of the PKA response cascade depends to a certain extent on which activation process is initiated (Ca^{2+}, PKC, CaN) and the targeting of PKA to synaptic proteins by AKAPs.

As noted above, PKA activity is important in the establishment of synaptic plasticity and the maintenance of altered synaptic transmission (Rosenmund et al., 1994; Brandon et al., 1995; Qi et al., 1996; Abel et al., 1997; de Lecea et al., 1998; Otmakhov and Lisman, 2002). An age-related deficit in maintenance of LTP is ameliorated by activation of PKA (Bach et al., 1999) suggesting that stimulated activation of PKA is reduced in aged animals during induction of LTP. Evidence has been provided that stimulated brain PKA activity is decreased in aged humans (Meier-Ruge et al., 1980), rodents (Karege et al., 2001b), and insects (Laviada et al., 1997). In addition, cAMP-induced protein phosphorylation of the PKA substrate, DARPP-32, is decreased with aging further supporting the idea that stimulation of PKA activity is reduced with age (Govoni et al., 1988). Moreover, when DARPP-32 is phosphorylated this molecule acts to inhibit protein phosphatase activity such that the net effect would appear to be a shift in the balance of kinases/phosphatases, favoring increased phosphatase activity. Indeed, a reduction in basal activity has been observed in aged rats; however, this reduction may be offset to a certain extent by a decrease in the level of the PKA regulatory subunit (Karege et al., 2001a). Nevertheless, a reduction in regulatory units would constrict the range of stimulated PKA activity and may underlie impair synaptic plasticity (Abel et al., 1997; de Lecea et al., 1998).

While much less is known about other kinases, there is some suggestion that expression of G protein-coupled receptor kinases decreases with age (Grange-Midroit et al., 2002), and the activation of tyrosine kinases and mitogen-activated protein (MAP) kinases may be reduced with advanced age (Mullany et al., 1996). For instance, while no overall change has been observed for the expression of SRC (Matocha et al., 1990), a tyrosine kinase family involved in synaptic plasticity (Soderling and Derkach, 2000), there appears to be a loss of receptors for activating tyrosine kinases (Sato et al., 2001; Romanczyk et al., 2002).

3.2. Increased phosphatase activity

The major serine/threonine protein phosphatases, PP1 and PP2A, as well as the Ca^{2+}-dependent phosphatase, CaN, are highly expressed in the brain (Price and Mumby, 1999). As noted earlier, inhibition of phosphatase activity selectively increased synaptic strength in aged animals suggesting that increased phosphatase may underlie the decrease in synaptic transmission during aging (Norris et al., 1998). Evidence from several different studies suggest that the increase in phosphatase activity is likely due to altered regulation of Ca^{2+} during aging. First, the expression of PP1, PP2A, and CaN do not appear to change with aging in humans and rodents (Foster et al., 2001; Grange-Midroit et al., 2002). Second, despite similar levels in whole hippocampal homogenates, CaN and PP1 activity were increased in the cytosolic fraction (Foster and Kumar, 2002; Foster et al., 2003). Moreover, the level of CaN, but not PP1, was increased in the cytosolic fraction. In this regard, it is important to note that CaN activates PP1 through dephosphorylation of the PP1 regulatory protein, inhibitor 1 or DARPP-32, such that an increase in CaN activity can drive increased PP1 activation. Finally, the hippocampus of aged animals exhibited reduced phosphorylation of CaN and PP1 substrate proteins. Interestingly, these proteins (BAD and CREB) are involved in apoptosis, memory consolidation, and regulation of synaptic strength (Mayford and Kandel, 1999; Wang et al., 1999; Walton and Dragunow, 2000).

The entry of Ca^{2+} into the cell initiates CaN activation, releasing CaN from the membrane into the cytosol. Indeed, the increase in CaN activity in the cytosol of aged animals appears to result from altered Ca^{2+} regulation, since incubation in a Ca^{2+}-free buffer, or application of Ca^{2+}-channel blockers 30 min prior to tissue collection, shifted CaN back to the membrane fraction (Foster and Kumar, 2002). Finally, studies involving transgenic animals support the idea that a shift in the balance of kinase/phosphatase activity influences the phosphorylation of synaptic proteins and memory function in a manner similar to that observed in aged animals (Mayford and Kandel, 1999; Genoux et al., 2002). Together, the results suggest that there is an increase in baseline phosphatase activity during aging, which might be expected to dephosphorylate proteins involved in synaptic transmission.

In contrast to normal aging, the hyperphosphorylation of tau protein in the brains of Alzheimer's patients may result from a decreased phosphatase activity (Gong et al., 1993, 1995). It is possible that the hyperphosphorylation is related to increased kinase activity due to altered Ca^{2+} homeostasis, which is further disrupted by the molecular changes associated with the disease (LaFerla, 2002). Alternatively, hyperphosphorylation may result from decreased phosphatase activity. A reduction in the expression of CaN has been observed in the tangle-bearing neurons (Pei et al., 1994), and the expression of mRNA for PP2A is reduced in Alzheimer's patients (Vogelsberg-Ragaglia et al., 2001).

4. Conclusion

In general, increases and decreases in hippocampal synaptic transmission are associated with increased activity of serine/threonine kinases and phosphatases, respectively. Thus far, research indicates that the decrease in synaptic transmission with aging is associated with a shift in the balance of phosphatase/kinase activity, favoring phosphatases. Overall, the deficit does not appear to be due to a change in enzyme expression, rather the difficulty appears to result from an impairment of mechanisms for the stimulation of kinase activity. Impairment in the activation of phosphorylation mechanisms would limit the functional plasticity of the presynaptic and postsynaptic elements. In addition, changes in Ca^{2+} regulation may promote increased basal CaN activity and the activation of phosphatase cascades involving PP1. While the findings suggest that a shift in enzyme activity underlies age-related changes in synaptic transmission and synaptic plasticity, the state of phosphoproteins involved in transmission is largely unknown. For the few studies that have examined phosphoproteins associated with the synapse, the results tend to support the notion of increased phosphatase over kinase activity.

Acknowledgments

This work was supported by the National Institutes of Health Grant AG14979 and the Evelyn F. McKnight Brain Research Grant.

References

Abdul-Ghani, M.A., Valiante, T.A., Carlen, P.L., Pennefather, P.S., 1996. Tyrosine kinase inhibitors enhance a Ca(2+)-activated K+ current (IAHP) and reduce IAHP suppression by a metabotropic glutamate receptor agonist in rat dentate granule neurones. J. Physiol. 496(Pt 1), 139–144.

Abel, T., Nguyen, P.V., Barad, M., Deuel, T.A., Kandel, E.R., Bourtchouladze, R., 1997. Genetic demonstration of a role for PKA in the late phase of LTP and in hippocampus-based long-term memory. Cell 88, 615–626.

Alagarsamy, S., Marino, M.J., Rouse, S.T., Gereau, R.W.T., Heinemann, S.F., Conn, P.J., 1999. Activation of NMDA receptors reverses desensitization of mGluR5 in native and recombinant systems. Nat. Neurosci. 2, 234–240.

Alaluf, S., Mulvihill, E.R., McIlhinney, R.A., 1995. Rapid agonist mediated phosphorylation of the metabotropic glutamate receptor 1 alpha by protein kinase C in permanently transfected BHK cells. FEBS Lett. 367, 301–305.

Antoni, F.A., Palkovits, M., Simpson, J., Smith, S.M., Leitch, A.L., Rosie, R., Fink, G., Paterson, J.M., 1998. Ca2+/calcineurin-inhibited adenylyl cyclase, highly abundant in forebrain regions, is important for learning and memory. J. Neurosci. 18, 9650–9661.

Anwyl, R., 1999. Metabotropic glutamate receptors: electrophysiological properties and role in plasticity. Brain Res. Brain Res. Rev. 29, 83–120.

Armbrecht, H.J., Boltz, M.A., Kumar, V.B., Flood, J.F., Morley, J.E., 1999. Effect of age on calcium-dependent proteins in hippocampus of senescence-accelerated mice. Brain Res. 842, 287–293.

Attucci, S., Clodfelter, G.V., Thibault, O., Staton, J., Moroni, F., Landfield, P.W., Porter, N.M., 2002. Group I metabotropic glutamate receptor inhibition selectively blocks a prolonged Ca(2+) elevation associated with age-dependent excitotoxicity. Neuroscience 112, 183–194.

Bach, M.E., Barad, M., Son, H., Zhuo, M., Lu, Y.F., Shih, R., Mansuy, I., Hawkins, R.D., Kandel, E.R., 1999. Age-related defects in spatial memory are correlated with defects in the late phase of hippocampal long-term potentiation in vitro and are attenuated by drugs that enhance the cAMP signaling pathway. Proc. Natl. Acad. Sci. USA 96, 5280–5285.

Banke, T.G., Bowie, D., Lee, H., Huganir, R.L., Schousboe, A., Traynelis, S.F., 2000. Control of GluR1 AMPA receptor function by cAMP-dependent protein kinase. J. Neurosci. 20, 89–102.

Barnes, C.A., Mizumori, S.J., Lovinger, D.M., Sheu, F.S., Murakami, K., Chan, S.Y., Linden, D.J., Nelson, R.B., Routtenberg, A., 1988. Selective decline in protein F1 phosphorylation in hippocampus of senescent rats. Neurobiol. Aging 9, 393–398.

Barnes, C.A., Rao, G., Orr, G., 2000. Age-related decrease in the Schaffer collateral-evoked EPSP in awake, freely behaving rats. Neural Plast. 7, 167–178.

Barria, A., Muller, D., Derkach, V., Griffith, L.C., Soderling, T.R., 1997. Regulatory phosphorylation of AMPA-type glutamate receptors by CaM-KII during long-term potentiation. Science 276, 2042–2045.

Bartschat, D.K., Rhodes, T.E., 1995. Protein kinase C modulates calcium channels in isolated presynaptic nerve terminals of rat hippocampus. J. Neurochem. 64, 2064–2072.

Battaini, F., Elkabes, S., Bergamaschi, S., Ladisa, V., Lucchi, L., De Graan, P.N., Schuurman, T., Wetsel, W.C., Trabucchi, M., Govoni, S., 1995. Protein kinase C activity, translocation, and conventional isoforms in aging rat brain. Neurobiol. Aging 16, 137–148.

Ben-Ari, Y., Aniksztejn, L., Bregestovski, P., 1992. Protein kinase C modulation of NMDA currents: an important link for LTP induction. Trends Neurosci. 15, 333–339.

Blackstone, C., Murphy, T.H., Moss, S.J., Baraban, J.M., Huganir, R.L., 1994. Cyclic AMP and synaptic activity-dependent phosphorylation of AMPA-preferring glutamate receptors. J. Neurosci. 14, 7585–7593.

Boxer, A.L., Moreno, H., Rudy, B., Ziff, E.B., 1999. FGF-2 potentiates Ca(2+)-dependent inactivation of NMDA receptor currents in hippocampal neurons. J. Neurophysiol. 82, 3367–3377.

Brandon, E.P., Zhuo, M., Huang, Y.Y., Qi, M., Gerhold, K.A., Burton, K.A., Kandel, E.R., McKnight, G.S., Idzerda, R.L., 1995. Hippocampal long-term depression and depotentiation are defective in mice carrying a targeted disruption of the gene encoding the RI beta subunit of cAMP-dependent protein kinase. Proc. Natl. Acad. Sci. USA 92, 8851–8855.

Brandon, E.P., Idzerda, R.L., McKnight, G.S., 1997. PKA isoforms, neural pathways, and behaviour: making the connection. Curr. Opin. Neurobiol. 7, 397–403.

Cai, Z., Saugstad, J.A., Sorensen, S.D., Ciombor, K.J., Zhang, C., Schaffhauser, H., Hubalek, F., Pohl, J., Duvoisin, R.M., Conn, P.J., 2001. Cyclic AMP-dependent protein kinase phosphorylates group III metabotropic glutamate receptors and inhibits their function as presynaptic receptors. J. Neurochem. 78, 756–766.

Carroll, R.C., Zukin, R.S., 2002. NMDA-receptor trafficking and targeting: implications for synaptic transmission and plasticity. Trends Neurosci. 25, 571–577.

Chen, L., Huang, L.Y., 1992. Protein kinase C reduces Mg2+ block of NMDA-receptor channels as a mechanism of modulation. Nature 356, 521–523.

Ciruela, F., Giacometti, A., McIlhinney, R.A., 1999. Functional regulation of metabotropic glutamate receptor type 1c: a role for phosphorylation in the desensitization of the receptor. FEBS Lett. 462, 278–282.

Coghlan, V.M., Perrino, B.A., Howard, M., Langeberg, L.K., Hicks, J.B., Gallatin, W.M., Scott, J.D., 1995. Association of protein kinase A and protein phosphatase 2B with a common anchoring protein. Science 267, 108–111.

Colledge, M., Dean, R.A., Scott, G.K., Langeberg, L.K., Huganir, R.L., Scott, J.D., 2000. Targeting of PKA to glutamate receptors through a MAGUK-AKAP complex. Neuron 27, 107–119.

Colombo, P.J., Wetsel, W.C., Gallagher, M., 1997. Spatial memory is related to hippocampal subcellular concentrations of calcium-dependent protein kinase C isoforms in young and aged rats. Proc. Natl. Acad. Sci. USA 94, 14195–14199.

Dale, L.B., Bhattacharya, M., Anborgh, P.H., Murdoch, B., Bhatia, M., Nakanishi, S., Ferguson, S.S., 2000. G protein-coupled receptor kinase-mediated desensitization of metabotropic glutamate receptor 1A protects against cell death. J. Biol. Chem. 275, 38213–38220.

Davis, S., Salin, H., Helme-Guizon, A., Dumas, S., Stephan, A., Corbex, M., Mallet, J., Laroche, S., 2000. Dysfunctional regulation of alphaCaMKII and syntaxin 1B transcription after induction of LTP in the aged rat. Eur. J. Neurosci. 12, 3276–3282.

de Lecea, L., Criado, J.R., Rivera, S., Wen, W., Soriano, E., Henriksen, S.J., Taylor, S.S., Gall, C.M., Sutcliffe, J.G., 1998. Endogenous protein kinase A inhibitor (PKIalpha) modulates synaptic activity. J. Neurosci. Res. 53, 269–278.

Dekker, L.V., De Graan, P.N., Gispen, W.H., 1991. Transmitter release: target of regulation by protein kinase C? Prog. Brain Res. 89, 209–233.

Derkach, V., Barria, A., Soderling, T.R., 1999. Ca2+/calmodulin-kinase II enhances channel conductance of alpha-amino-3-hydroxy-5-methyl-4-isoxazolepropionate type glutamate receptors. Proc. Natl. Acad. Sci. USA 96, 3269–3274.

Dingledine, R., Borges, K., Bowie, D., Traynelis, S.F., 1999. The glutamate receptor ion channels. Pharmacol. Rev. 51, 7–61.

Douma, B.R., Van der Zee, E.A., Luiten, P.G., 1998. Translocation of protein kinase Cgamma occurs during the early phase of acquisition of food rewarded spatial learning. Behav. Neurosci. 112, 496–501.

Eckles, K.E., Dudek, E.M., Bickford, P.C., Browning, M.D., 1997. Amelioration of age-related deficits in the stimulation of synapsin phosphorylation. Neurobiol. Aging 18, 213–217.

Ehlers, M.D., 2000. Reinsertion or degradation of AMPA receptors determined by activity-dependent endocytic sorting. Neuron 28, 511–525.

Ehlers, M.D., Tingley, W.G., Huganir, R.L., 1995. Regulated subcellular distribution of the NR1 subunit of the NMDA receptor. Science 269, 1734–1737.

Evans, D.I., Jones, R.S., Woodhall, G., 2001. Differential actions of PKA and PKC in the regulation of glutamate release by group III mGluRs in the entorhinal cortex. J. Neurophysiol. 85, 571–579.

Ferreira, A., Rapoport, M., 2002. The synapsins: beyond the regulation of neurotransmitter release. Cell. Mol. Life Sci. 59, 589–595.

Fisher, E.H., Kreb, E.G., 1955. Conversion of phosphorylase b to phosphorylase a in muscle extracts. J. Biol. Chem. 216, 121–132.

Fordyce, D.E., Wehner, J.M., 1993. Effects of aging on spatial learning and hippocampal protein kinase C in mice. Neurobiol. Aging 14, 309–317.

Foster, T.C., 1999. Involvement of hippocampal synaptic plasticity in age-related memory decline. Brain Res. Rev. 30, 236–249.

Foster, T.C., 2002. Regulation of synaptic plasticity in memory and memory decline with aging. Prog. Brain Res. 138, 283–303.

Foster, T.C., Norris, C.M., 1997. Age-associated changes in Ca(2+)-dependent processes: relation to hippocampal synaptic plasticity. Hippocampus 7, 602–612.

Foster, T.C., Kumar, A., 2002. Calcium dysregulation in the aging brain. Neuroscientist 8, 297–301.

Foster, T.C., Sharrow, K.M., Masse, J.R., Norris, C.M., Kumar, A., 2001. Calcineurin links Ca2+ dysregulation with brain aging. J. Neurosci. 21, 4066–4073.

Foster, T.C., Sharrow, K.M., Kumar, A., Masse, J., 2003. Interaction of age and chronic estradiol replacement on memory and markers of brain aging. Neurobiol. Aging 24, 839–852.

Friedman, E., Wang, H.Y., 1989. Effect of age on brain cortical protein kinase C and its mediation of 5-hydroxytryptamine release. J. Neurochem. 52, 187–192.

Gao, J., Yin, D., Yao, Y., Williams, T.D., Squier, T.C., 1998. Progressive decline in the ability of calmodulin isolated from aged brain to activate the plasma membrane Ca-ATPase. Biochemistry 37, 9536–9548.

Genoux, D., Haditsch, U., Knobloch, M., Michalon, A., Storm, D., Mansuy, I.M., 2002. Protein phosphatase 1 is a molecular constraint on learning and memory. Nature 418, 970–975.

Gereau, R.W.T., Heinemann, S.F., 1998. Role of protein kinase C phosphorylation in rapid desensitization of metabotropic glutamate receptor 5. Neuron 20, 143–151.

Ghetti, A., Heinemann, S.F., 2000. NMDA-dependent modulation of hippocampal kainate receptors by calcineurin and Ca(2 +)/calmodulin-dependent protein kinase. J. Neurosci. 20, 2766–2773.

Gianotti, C., Porta, A., De Graan, P.N., Oestreicher, A.B., Nunzi, M.G., 1993. B-50/GAP-43 phosphorylation in hippocampal slices from aged rats: effects of phosphatidylserine administration. Neurobiol. Aging 14, 401–406.

Gong, C.X., Singh, T.J., Grundke-Iqbal, I., Iqbal, K., 1993. Phosphoprotein phosphatase activities in Alzheimer disease brain. J. Neurochem. 61, 921–927.

Gong, C.X., Shaikh, S., Wang, J.Z., Zaidi, T., Grundke-Iqbal, I., Iqbal, K., 1995. Phosphatase activity toward abnormally phosphorylated tau: decrease in Alzheimer disease brain. J. Neurochem. 65, 732–738.

Govoni, S., Rius, R.A., Battaini, F., Trabucchi, M., 1988. Reduced cAMP-dependent phosphorylation in striatum and nucleus accumbens of aged rats: evidence of an altered functioning of D1 dopaminoceptive neurons. J. Gerontol. 43, B93–B97.

Graef, I.A., Mermelstein, P.G., Stankunas, K., Neilson, J.R., Deisseroth, K., Tsien, R.W., Crabtree, G.R., 1999. L-type calcium channels and GSK-3 regulate the activity of NF-ATc4 in hippocampal neurons. Nature 401, 703–708.

Grange-Midroit, M., Garcia-Sevilla, J.A., Ferrer-Alcon, M., La Harpe, R., Walzer, C., Guimon, J., 2002. G protein-coupled receptor kinases, beta-arrestin-2 and associated regulatory proteins in the human brain: postmortem changes, effect of age and subcellular distribution. Brain Res. Mol. Brain Res. 101, 39–51.

Greengard, P., Jen, J., Nairn, A.C., Stevens, C.F., 1991. Enhancement of the glutamate response by cAMP-dependent protein kinase in hippocampal neurons. Science 253, 1135–1138.

Gu, Q., Moss, R.L., 1996. 17 beta-Estradiol potentiates kainate-induced currents via activation of the cAMP cascade. J. Neurosci. 16, 3620–3629.

He, X., Yang, F., Xie, Z., Lu, B., 2000. Intracellular Ca(2 +) and Ca(2 +)/calmodulin-dependent kinase II mediate acute potentiation of neurotransmitter release by neurotrophin-3. J. Cell Biol. 149, 783–792.

Heidinger, V., Manzerra, P., Wang, X.Q., Strasser, U., Yu, S.P., Choi, D.W., Behrens, M.M., 2002. Metabotropic glutamate receptor 1-induced upregulation of NMDA receptor current: mediation through the Pyk2/Src-family kinase pathway in cortical neurons. J. Neurosci. 22, 5452–5461.

Hell, J.W., Yokoyama, C.T., Breeze, L.J., Chavkin, C., Catterall, W.A., 1995. Phosphorylation of presynaptic and postsynaptic calcium channels by cAMP-dependent protein kinase in hippocampal neurons. EMBO J. 14, 3036–3044.

Hilfiker, S., Pieribone, V.A., Czernik, A.J., Kao, H.T., Augustine, G.J., Greengard, P., 1999. Synapsins as regulators of neurotransmitter release. Phil. Trans. R. Soc. Lond. B. Biol. Sci. 354, 269–279.

Hilfiker, S., Czernik, A.J., Greengard, P., Augustine, G.J., 2001. Tonically active protein kinase A regulates neurotransmitter release at the squid giant synapse. J. Physiol. 531, 141–146.

Hodgkiss, J.P., Kelly, J.S., 1995. Only "de novo" long-term depression (LTD) in the rat hippocampus in vitro is blocked by the same low concentration of FK506 that blocks LTD in the visual cortex. Brain Res. 705, 241–246.

Hrabetova, S., Sacktor, T.C., 2001. Transient translocation of conventional protein kinase C isoforms and persistent downregulation of atypical protein kinase Mzeta in long-term depression. Brain Res. Mol. Brain Res. 95, 146–152.

Huang, C.C., Liang, Y.C., Hsu, K.S., 2001. Characterization of the mechanism underlying the reversal of long term potentiation by low frequency stimulation at hippocampal CA1 synapses. J. Biol. Chem. 276, 48108–48117.

Hudmon, A., Schulman, H., 2002. Neuronal CA2 + /calmodulin-dependent protein kinase II: the role of structure and autoregulation in cellular function. Annu. Rev. Biochem. 71, 473–510.

Jouvenceau, A., Dutar, P., Billard, J.M., 1997. Is activation of the metabotropic glutamate receptors impaired in the hippocampal CA1 area of the aged rat? Hippocampus 7, 455–459.

Kameyama, K., Lee, H.K., Bear, M.F., Huganir, R.L., 1998. Involvement of a postsynaptic protein kinase A substrate in the expression of homosynaptic long-term depression. Neuron 21, 1163–1175.

Karege, F., Lambercy, C., Schwald, M., Steimer, T., Cisse, M., 2001a. Differential changes of cAMP-dependent protein kinase activity and 3H-cAMP binding sites in rat hippocampus during maturation and aging. Neurosci. Lett. 315, 89–92.

Karege, F., Schwald, M., Lambercy, C., Murama, J.J., Cisse, M., Malafosse, A., 2001b. A non-radioactive assay for the cAMP-dependent protein kinase activity in rat brain homogenates and age-related changes in hippocampus and cortex. Brain Res. 903, 86–93.

La Porta, C.A., Comolli, R., 1999. Age-dependent modulation of PKC isoforms and NOS activity and expression in rat cortex, striatum, and hippocampus. Exp. Gerontol. 34, 863–874.

LaFerla, F.M., 2002. Calcium dyshomeostasis and intracellular signalling in Alzheimer's disease. Nat. Rev. Neurosci. 3, 862–872.

Lan, J.Y., Skeberdis, V.A., Jover, T., Grooms, S.Y., Lin, Y., Araneda, R.C., Zheng, X., Bennett, M.V., Zukin, R.S., 2001. Protein kinase C modulates NMDA receptor trafficking and gating. Nat. Neurosci. 4, 382–390.

Laviada, I.D., Galve-Roperh, I., Malpartida, J.M., Haro, A., 1997. cAMP signalling mechanisms with aging in the Ceratitis capitata brain. Mech. Ageing Dev. 97, 45–53.

Lee, H.K., Kameyama, K., Huganir, R.L., Bear, M.F., 1998. NMDA induces long-term synaptic depression and dephosphorylation of the GluR1 subunit of AMPA receptors in hippocampus. Neuron 21, 1151–1162.

Lee, H.K., Barbarosie, M., Kameyama, K., Bear, M.F., Huganir, R.L., 2000. Regulation of distinct AMPA receptor phosphorylation sites during bidirectional synaptic plasticity. Nature 405, 955–959.

Li, B., Chen, N., Luo, T., Otsu, Y., Murphy, T.H., Raymond, L.A., 2002. Differential regulation of synaptic and extra-synaptic NMDA receptors. Nat. Neurosci. 5, 833–834.

Liao, G.Y., Wagner, D.A., Hsu, M.H., Leonard, J.P., 2001. Evidence for direct protein kinase-C mediated modulation of N-methyl-D-aspartate receptor current. Mol. Pharmacol. 59, 960–964.

Lieberman, D.N., Mody, I., 1994. Regulation of NMDA channel function by endogenous Ca(2+)-dependent phosphatase. Nature 369, 235–239.

Lin, M.J., Lin-Shiau, S.Y., 1999. Enhanced spontaneous transmitter release at murine motor nerve terminals with cyclosporine. Neuropharmacology 38, 195–198.

Lin, R.C., Scheller, R.H., 2000. Mechanisms of synaptic vesicle exocytosis. Annu. Rev. Cell Dev. Biol. 16, 19–49.

Lin, J.W., Ju, W., Foster, K., Lee, S.H., Ahmadian, G., Wyszynski, M., Wang, Y.T., Sheng, M., 2000. Distinct molecular mechanisms and divergent endocytotic pathways of AMPA receptor internalization. Nat. Neurosci. 3, 1282–1290.

Magnusson, K.R., 1998. Aging of glutamate receptors: correlations between binding and spatial memory performance in mice. Mech. Ageing Dev. 104, 227–248.

Mammen, A.L., Kameyama, K., Roche, K.W., Huganir, R.L., 1997. Phosphorylation of the alpha-amino-3-hydroxy-5-methylisoxazole-4-propionic acid receptor GluR1 subunit by calcium/calmodulin-dependent kinase II. J. Biol. Chem. 272, 32528–32533.

Markram, H., Segal, M., 1992. Activation of protein kinase C suppresses responses to NMDA in rat CA1 hippocampal neurones. J. Physiol. 457, 491–501.

Matocha, M.F., Fitzpatrick, S.W., Atack, J.R., Rapoport, S.I., 1990. pp60c-src kinase expression in brain of adult rats in relation to age. Exp. Gerontol. 25, 47–54.

Mayford, M., Kandel, E.R., 1999. Genetic approaches to memory storage. Trends Genet. 15, 463–470.

Mei, Y., Gawai, K.R., Nie, Z., Ramkumar, V., Helfert, R.H., 1999. Age-related reductions in the activities of antioxidant enzymes in the rat inferior colliculus. Hear. Res. 135, 169–180.

Meier-Ruge, W., Iwangoff, P., Reichlmeier, K., Sandoz, P., 1980. Neurochemical findings in the aging brain. Adv. Biochem. Psychopharmacol. 23, 323–338.

Meyer, E.M., Judkins, J.H., 1993. The development of age-related deficits in several presynaptic processes associated with brain [3H]acetylcholine release. Mech. Ageing Dev. 72, 119–128.

Meyer, E.M., Judkins, J.H., Momol, A.E., Hardwick, E.O., 1994. Effects of peroxidation and aging on rat neocortical ACh-release and protein kinase C. Neurobiol. Aging 15, 63–67.

　　　　　　　　　　　　　　　　　　　　T. C. Foster

Minakami, R., Jinnai, N., Sugiyama, H., 1997. Phosphorylation and calmodulin binding of the metabotropic glutamate receptor subtype 5 (mGluR5) are antagonistic in vitro. J. Biol. Chem. 272, 20291–20298.

Mizutani, T., Nakashima, S., Nozawa, Y., 1998. Changes in the expression of protein kinase C (PKC), phospholipases C (PLC) and D (PLD) isoforms in spleen, brain and kidney of the aged rat: RT-PCR and Western blot analysis. Mech. Ageing Dev. 105, 151–172.

Mons, N., Enderlin, V., Jaffard, R., Higueret, P., 2001. Selective age-related changes in the PKC-sensitive, calmodulin-binding protein, neurogranin, in the mouse brain. J. Neurochem. 79, 859–867.

Mulkey, R.M., Endo, S., Shenolikar, S., Malenka, R.C., 1994. Involvement of a calcineurin/inhibitor-1 phosphatase cascade in hippocampal long-term depression. Nature 369, 486–488.

Mullany, P., Connolly, S., Lynch, M.A., 1996. Ageing is associated with changes in glutamate release, protein tyrosine kinase and Ca2+/calmodulin-dependent protein kinase II in rat hippocampus. Eur. J. Pharmacol. 309, 311–315.

Nicolle, M.M., Colombo, P.J., Gallagher, M., McKinney, M., 1999. Metabotropic glutamate receptor-mediated hippocampal phosphoinositide turnover is blunted in spatial learning-impaired aged rats. J. Neurosci. 19, 9604–9610.

Norris, C.M., Halpain, S., Foster, T.C., 1998. Alterations in the balance of protein kinase/phosphatase activities parallel reduced synaptic strength during aging. J. Neurophysiol. 80, 1567–1570.

Norris, C.M., Blalock, E.M., Chen, K.C., Porter, N.M., Landfield, P.W., 2002. Calcineurin enhances L-type Ca(2+) channel activity in hippocampal neurons: increased effect with age in culture. Neuroscience 110, 213–225.

Orlando, L.R., Dunah, A.W., Standaert, D.G., Young, A.B., 2002. Tyrosine phosphorylation of the metabotropic glutamate receptor mGluR5 in striatal neurons. Neuropharmacology 43, 161–173.

Otmakhov, N., Lisman, J.E., 2002. Postsynaptic application of a cAMP analogue reverses long-term potentiation in hippocampal CA1 pyramidal neurons. J. Neurophysiol. 87, 3018–3032.

Parent, A., Rowe, W., Meaney, M.J., Quirion, R., 1995. Increased production of inositol phosphates and diacylglycerol in aged cognitively impaired rats after stimulation of muscarinic, metabotropic-glutamate and endothelin receptors. J. Pharmacol. Exp. Ther. 272, 1110–1116.

Parfitt, K.D., Hoffer, B.J., Browning, M.D., 1991. Norepinephrine and isoproterenol increase the phosphorylation of synapsin I and synapsin II in dentate slices of young but not aged Fisher 344 rats. Proc. Natl. Acad. Sci. USA 88, 2361–2365.

Pascale, A., Govoni, S., Battaini, F., 1998a. Age-related alteration of PKC, a key enzyme in memory processes: physiological and pathological examples. Mol. Neurobiol. 16, 49–62.

Pascale, A., Nogues, X., Marighetto, A., Micheau, J., Battaini, F., Govoni, S., Jaffard, R., 1998b. Cytosolic hippocampal PKC and aging: correlation with discrimination performance. Neuroreport 9, 725–729.

Pei, J.J., Sersen, E., Iqbal, K., Grundke-Iqbal, I., 1994. Expression of protein phosphatases (PP-1, PP-2A, PP-2B and PTP-1B) and protein kinases (MAP kinase and P34cdc2) in the hippocampus of patients with Alzheimer disease and normal aged individuals. Brain Res. 655, 70–76.

Pintor, A., Tiburzi, F., Pezzola, A., Volpe, M.T., 1998. Metabotropic glutamate receptor agonist (1S,3R-ACPD) increased frontal cortex dopamine release in aged but not in young rats. Eur. J. Pharmacol. 359, 139–142.

Pintor, A., Potenza, R.L., Domenici, M.R., Tiburzi, F., Reggio, R., Pezzola, A., Popoli, P., 2000. Age-related decline in the functional response of striatal group I mGlu receptors. Neuroreport 11, 3033–3038.

Price, N.E., Mumby, M.C., 1999. Brain protein serine/threonine phosphatases. Curr. Opin. Neurobiol. 9, 336–342.

Qi, M., Zhuo, M., Skalhegg, B.S., Brandon, E.P., Kandel, E.R., McKnight, G.S., Idzerda, R.L., 1996. Impaired hippocampal plasticity in mice lacking the Cbeta1 catalytic subunit of cAMP-dependent protein kinase. Proc. Natl. Acad. Sci. USA 93, 1571–1576.

Racchi, M., Govoni, S., Solerte, S.B., Galli, C.L., Corsini, E., 2001. Dehydroepiandrosterone and the relationship with aging and memory: a possible link with protein kinase C functional machinery. Brain Res. Brain Res. Rev. 37, 287–293.

Rahman, S., McLean, J.H., Darby-King, A., Paterno, G., Reynolds, J.N., Neuman, R.S., 1995. Loss of cortical serotonin2A signal transduction in senescent rats: reversal following inhibition of protein kinase C. Neuroscience 66, 891–901.

Ramakers, G.M., Pasinelli, P., Hens, J.J., Gispen, W.H., De Graan, P.N., 1997. Protein kinase C in synaptic plasticity: changes in the in situ phosphorylation state of identified pre- and postsynaptic substrates. Prog. Neuropsychopharmacol. Biol. Psychiatry 21, 455–486.

Ramakers, G.M., Heinen, K., Gispen, W.H., de Graan, P.N., 2000. Long term depression in the CA1 field is associated with a transient decrease in pre- and postsynaptic PKC substrate phosphorylation. J. Biol. Chem. 275, 28682–28687.

Raman, I.M., Tong, G., Jahr, C.E., 1996. Beta-adrenergic regulation of synaptic NMDA receptors by cAMP-dependent protein kinase. Neuron 16, 415–421.

Raymond, L.A., Blackstone, C.D., Huganir, R.L., 1993. Phosphorylation and modulation of recombinant GluR6 glutamate receptors by cAMP dependent protein kinase. Nature 361, 637–641.

Robinson, P.J., 1991. The role of protein kinase C and its neuronal substrates dephosphin, B-50, and MARCKS in neurotransmitter release. Mol. Neurobiol. 5, 87–130.

Roche, K.W., O'Brien, R.J., Mammen, A.L., Bernhardt, J., Huganir, R.L., 1996. Characterization of multiple phosphorylation sites on the AMPA receptor GluR1 subunit. Neuron 16, 1179–1188.

Roche, K.W., Standley, S., McCallum, J., Dune Ly, C., Ehlers, M.D., Wenthold, R.J., 2001. Molecular determinants of NMDA receptor internalization. Nat. Neurosci. 4, 794–802.

Roeper, J., Lorra, C., Pongs, O., 1997. Frequency-dependent inactivation of mammalian A-type K + channel KV1.4 regulated by Ca2 + /calmodulin-dependent protein kinase. J. Neurosci. 17, 3379–3391.

Romanczyk, T.B., Weickert, C.S., Webster, M.J., Herman, M.M., Akil, M., Kleinman, J.E., 2002. Alterations in trkB mRNA in the human prefrontal cortex throughout the lifespan. Eur. J. Neurosci. 15, 269–280.

Rosenmund, C., Carr, D.W., Bergeson, S.E., Nilaver, G., Scott, J.D., Westbrook, G.L., 1994. Anchoring of protein kinase A is required for modulation of AMPA/kainate receptors on hippocampal neurons. Nature 368, 853–856.

Sato, T., Wilson, T.S., Hughes, L.F., Konrad, H.R., Nakayama, M., Helfert, R.H., 2001. Age-related changes in levels of tyrosine kinase B receptor and fibroblast growth factor receptor 2 in the rat inferior colliculus: implications for neural senescence. Neuroscience 103, 695–702.

Schaffhauser, H., Cai, Z., Hubalek, F., Macek, T.A., Pohl, J., Murphy, T.J., Conn, P.J., 2000. cAMP-dependent protein kinase inhibits mGluR2 coupling to G-proteins by direct receptor phosphorylation. J. Neurosci. 20, 5663–5670.

Scott, D.B., Blanpied, T.A., Swanson, G.T., Zhang, C., Ehlers, M.D., 2001. An NMDA receptor ER retention signal regulated by phosphorylation and alternative splicing. J. Neurosci. 21, 3063–3072.

Segovia, G., Porras, A., Del Arco, A., Mora, F., 2001. Glutamatergic neurotransmission in aging: a critical perspective. Mech. Ageing Dev. 122, 1–29.

Sim, A.T., Lloyd, H.G., Jarvie, P.E., Morrison, M., Rostas, J.A., Dunkley, P.R., 1993. Synaptosomal amino acid release: effect of inhibiting protein phosphatases with okadaic acid. Neurosci. Lett. 160, 181–184.

Simonyi, A., Miller, L.A., Sun, G.Y., 2000. Region-specific decline in the expression of metabotropic glutamate receptor 7 mRNA in rat brain during aging. Brain Res. Mol. Brain Res. 82, 101–106.

Soderling, T.R., Derkach, V.A., 2000. Postsynaptic protein phosphorylation and LTP. Trends Neurosci. 23, 75–80.

Song, I., Huganir, R.L., 2002. Regulation of AMPA receptors during synaptic plasticity. Trends Neurosci. 25, 578–588.

Staak, S., Behnisch, T., Angenstein, F., 1995. Hippocampal long-term potentiation: transient increase but no persistent translocation of protein kinase C isoenzymes alpha and beta. Brain Res. 682, 55–62.

Svenningsson, P., Tzavara, E.T., Witkin, J.M., Fienberg, A.A., Nomikos, G.G., Greengard, P., 2002. Involvement of striatal and extrastriatal DARPP-32 in biochemical and behavioral effects of fluoxetine (Prozac). Proc. Natl. Acad. Sci. USA 99, 3182–3187.

Swope, S.L., Moss, S.I., Raymond, L.A., Huganir, R.L., 1999. Regulation of ligand-gated ion channels by protein phosphorylation. Adv. Second Messenger Phosphoprotein Res. 33, 49–78.

Sze, C., Bi, H., Kleinschmidt-DeMasters, B.K., Filley, C.M., Martin, L.J., 2001. *N*-Methyl-D-aspartate receptor subunit proteins and their phosphorylation status are altered selectively in Alzheimer's disease. J. Neurol. Sci. 182, 151–159.

Tingley, W.G., Ehlers, M.D., Kameyama, K., Doherty, C., Ptak, J.B., Riley, C.T., Huganir, R.L., 1997. Characterization of protein kinase A and protein kinase C phosphorylation of the *N*-methyl-D-aspartate receptor NR1 subunit using phosphorylation site-specific antibodies. J. Biol. Chem. 272, 5157–5166.

Tokuda, M., Hatase, O., 1998. Regulation of neuronal plasticity in the central nervous system by phosphorylation and dephosphorylation. Mol. Neurobiol. 17, 137–156.

Tomizawa, K., Ohta, J., Matsushita, M., Moriwaki, A., Li, S.T., Takei, K., Matsui, H., 2002. Cdk5/p35 regulates neurotransmitter release through phosphorylation and downregulation of P/Q-type voltage-dependent calcium channel activity. J. Neurosci. 22, 2590–2597.

Torii, N., Kamishita, T., Otsu, Y., Tsumoto, T., 1995. An inhibitor for calcineurin, FK506, blocks induction of long-term depression in rat visual cortex. Neurosci. Lett. 185, 1–4.

Undie, A.S., Wang, H.Y., Friedman, E., 1995. Decreased phospholipase C-beta immunoreactivity, phosphoinositide metabolism, and protein kinase C activation in senescent F-344 rat brain. Neurobiol. Aging 16, 19–28.

Unnerstall, J.R., Ladner, A., 1994. Deficits in the activation and phosphorylation of hippocampal tyrosine hydroxylase in the aged Fischer 344 rat following intraventricular administration of 6-hydroxy-dopamine. J. Neurochem. 63, 280–290.

Van der Zee, E.A., Luiten, P.G., Disterhoft, J.F., 1997. Learning-induced alterations in hippocampal PKC-immunoreactivity: a review and hypothesis of its functional significance. Prog. Neuropsychopharmacol. Biol. Psychiatry 21, 531–572.

Verona, M., Zanotti, S., Schafer, T., Racagni, G., Popoli, M., 2000. Changes of synaptotagmin interaction with t-SNARE proteins in vitro after calcium/calmodulin-dependent phosphorylation. J. Neurochem. 74, 209–221.

Vickroy, T.W., Malphurs, W.L., Carriger, M.L., 1995. Regulation of stimulus-dependent hippocampal acetylcholine release by okadaic acid-sensitive phosphoprotein phosphatases. Neurosci. Lett. 191, 200–204.

Vissel, B., Krupp, J.J., Heinemann, S.F., Westbrook, G.L., 2001. A use-dependent tyrosine dephosphorylation of NMDA receptors is independent of ion flux. Nat. Neurosci. 4, 587–596.

Vogelsberg-Ragaglia, V., Schuck, T., Trojanowski, J.Q., Lee, V.M., 2001. PP2A mRNA expression is quantitatively decreased in Alzheimer's disease hippocampus. Exp. Neurol. 168, 402–412.

Walton, M.R., Dragunow, I., 2000. Is CREB a key to neuronal survival? Trends Neurosci. 23, 48–53.

Wang, L.Y., Salter, M.W., MacDonald, J.F., 1991. Regulation of kainate receptors by cAMP-dependent protein kinase and phosphatases. Science 253, 1132–1135.

Wang, L.Y., Orser, B.A., Brautigan, D.L., MacDonald, J.F., 1994. Regulation of NMDA receptors in cultured hippocampal neurons by protein phosphatases 1 and 2A. Nature 369, 230–232.

Wang, Y.T., Yu, X.M., Salter, M.W., 1996. Ca(2+)-independent reduction of *N*-methyl-D-aspartate channel activity by protein tyrosine phosphatase. Proc. Natl. Acad. Sci. USA 93, 1721–1725.

Wang, H.G., Pathan, N., Ethell, I.M., Krajewski, S., Yamaguchi, Y., Shibasaki, F., McKeon, F., Bobo, T., Franke, T.F., Reed, J.C., 1999. Ca2+-induced apoptosis through calcineurin dephosphorylation of BAD. Science 284, 339–343.

Watson, J.B., Khorasani, H., Persson, A., Huang, K.P., Huang, F.L., O'Dell, T.J., 2002. Age-related deficits in long-term potentiation are insensitive to hydrogen peroxide: coincidence with enhanced autophosphorylation of Ca2+/calmodulin-dependent protein kinase II. J. Neurosci. Res. 70, 298–308.

Yakel, J.L., 1997. Calcineurin regulation of synaptic function: from ion channels to transmitter release and gene transcription. Trends Pharmacol. Sci. 18, 124–134.

**Advances in
Cell Aging and
Gerontology**

Animal models of tau phosphorylation and tauopathy – what have they taught us?

Lit-Fui Lau and Joel B. Schachter

*CNS Discovery, Pfizer Global Research and Development, Groton, CT 06340, USA.
Correspondence address: Lit-Fui Lau, Ph.D., MS220-4013, CNS Discovery,
Pfizer Global Research and Development, Groton, CT 06340, USA.
Tel.: + 1-860-715-1921; fax: + 1-860-715-2349.
E-mail address: lit-fui_lau@groton.pfizer.com*

Contents

Abstract. Alzheimer's disease (AD) is characterized by the presence of senile plaques and neurofibrillary tangles (NFTs) in brain. NFTs are mainly composed of aggregates of hyper-phosphorylated tau proteins. Various animal models have been generated that recapitulate different aspects of tau pathologies, including tau hyperphosphorylation, somatodendritic accumulation of tau in the pretangle stage, and deposition of NFTs. Amyloid peptides appear to potentiate but not initiate NFT deposition in transgenic mice. Tau phosphorylation and aging in various animal models also promote NFT deposition. In addition to tau pathologies, axonal degeneration, neuronal cell death, and/or behavioral deficits are apparent in a number of animal models. While mechanisms leading to neurodegeneration are believed to involve func-tional disruption of microtubules by either overexpression or hyperphosphorylation of tau, the formation of NFTs may confer additional toxicities. Although the present animal models of tau phosphorylation and tauopathy do not fully recapitulate AD pathologies, they provide significant insights into mechanisms leading to tau pathology and associated neurodegeneration, and may serve as *in vivo* models for testing potential drug candidates for the treatment of this debilitating disease.

Advances in Cell Aging and Gerontology, vol. 16, 153–176
DOI: 10.1016/S1566-3124(04)16007-0

1. Tauopathy in neurodegenerative diseases

Age-related neurodegenerative disorders afflict a significant and rapidly growing population of people throughout the world. Nearly 15% of individuals over 65 years old display some form of dementia and this number increases to over 40% by 80 years (Kawas and Katzman, 1999). About two-third of the dementia cases are diagnosed as Alzheimer's disease (AD), making it the most prominent form of dementia. Current estimates suggest that 18 million people are afflicted with AD worldwide and this number is expected to double over the next 30 years (Spillantini and Goedert, 1998; Bosanquet, 2001). Apart from the morbidity of the disease progression, the healthcare-related costs of AD are estimated at over $100 billion per year in the US alone (Johnson et al., 2000). Therapeutic interventions that could reduce disease severity or delay its progression could have dramatic socio-economic impact.

Pathological hallmarks of AD include β-amyloid (Aβ)-containing plaques (Selkoe, 2001) and neurofibrillary tangles (NFTs) composed of hyperphosphorylated tau (Trojanowski and Lee, 2000). In less than 5% of AD cases, the disease cosegregates with mutations of the amyloid precursor protein (APP) or one of two presenilin genes (PS1 or PS2) (Finch and Tanzi, 1997). The demonstration that these mutations increase the formation of amyloidogenic Aβ peptides in cellular models (Citron et al., 1992; Cai et al., 1993; Suzuki et al., 1994) is consistent with a proposed role for Aβ overproduction in the pathogenesis of this inherited form of familial Alzheimer's disease (FAD). Consequently, various drug discovery strategies have focused on blocking the secretase enzymes involved in producing Aβ from APP, on preventing the aggregation of Aβ into fibrils, and on promoting clearance of amyloid plaques. Though recognized as a disease-mediated pathology, the regulation of NFT formation was not immediately embraced as a therapeutic target for AD. This was due, in part, to the perception that the events leading to NFT formation were amongst the many downstream consequences of the neurodegenerative process, rather than being part of the cause. The recent identification of a series of tau mutations associated with frontotemporal dementias (Hutton et al., 1998; Poorkaj et al., 1998; Spillantini and Goedert, 1998) and the demonstration that these tau mutations lead to NFT formation and neurodegeneration in transgenic mice have changed the way that tau function and NFT formation are viewed in the neurodegenerative process. Although tau mutations per se are not observed in AD, the recognition that changes in the normal properties of tau are sufficient to cause neurodegeneration has established new interest in tau as a therapeutic target for AD.

Etiological studies of AD pathologies have often selectively focused on the individual roles of plaques or of tangles in the disease process, giving rise to two separate "religions" in the research community (Lee, 2001; Mudher and Lovestone, 2002). The βaptists posit that errors in normal APP processing are among the earliest events in the initiation of a neurodegenerative cascade and therefore play a primary causal role. Supporters of this view cite genetic evidence from FAD cases, linking APP and presenilin mutations to Aβ production, plaque formation, and genetic penetrance of the disease. Additionally, the presence of dystrophic neurites surrounding amyloid plaques in AD brain and the toxicity of fibrillar $A\beta$ to cultured neurons are also cited to support the view that fibrillar $A\beta$ is the mediator of the neurodegenerative process in AD (Hardy and Selkoe, 2002). In contrast to βaptist doctrine, the Tauists point to data indicating that abundant $A\beta$ deposits have been found in the brains of cognitively normal, aged individuals, whereas neurofibrillary degeneration correlates better with cognitive decline (Arriagada et al., 1992; Nagy et al., 1995; Berg et al., 1998; Giannakopoulos et al., 2000). Transgenic mice expressing mutant PS1 or doubly transgenic animals expressing mutant PS1 and mutant APP proteins display age-dependent amyloid plaque deposition in the absence of NFT formation, yet these animals do not demonstrate behavioral deficits that correlate with either the induction of plaque deposition or the density of plaque burden (Holcomb et al., 1999). Furthermore, several other neurodegenerative conditions, such as dementia pugilistica, frontotemporal dementia and Parkinsonism linked to chromosome-17 (FTDP-17), progressive supranuclear palsy, and cortico-basal degeneration, demonstrate an association of dementia with neurofibrillary degeneration in the absence of β-amyloid plaques (Geddes et al., 1994; Buée et al., 2000; Albers and Augood, 2001; Lee et al., 2001). Although the relationships between plaques and tangles, and their relative roles in AD progression, have not been definitively established, studies in cultured neurons indicate that $A\beta$ fibrils can activate cellular-signaling mechanisms that cause hyperphosphorylation of tau (Takashima et al., 1998; Alvarez et al., 2002; Zheng et al., 2002).

In vitro studies suggest that hyperphosphorylation of tau disrupts its normal function of binding to, and stabilizing the structure of, microtubules (Alonso et al., 1994). Hyperphosphorylation thus releases tau from microtubule and promotes aggregation of tau into multimeric structures (Alonso et al., 1996, 1997). Phosphorylated tau can self-assemble *in vitro* into straight filaments (SFs) and paired helical filament (PHF) structures, like those that compose NFT (Alonso et al., 2001). Enzymatic dephosphorylation of PHF from AD brain releases tau from the insoluble aggregate and restores its ability to promote microtubule assembly (Iqbal et al., 1994; Wang et al., 1995). The complete mechanism of NFT formation in AD brain remains to be established, but at least a portion of the enhanced phosphorylation of tau is thought to precede and promote PHF formation (Braak et al., 1994; Braak and Braak, 1998). A key question is whether polymeric tau filaments are directly responsible for promoting neurodegeneration or whether the dissociation of tau from microtubules leads to cytotoxic microtubule dysfunction, with tau aggregates forming as a late-stage marker in slowly degenerating neurons. In support of the view that dysregulation of normal tau functions is responsible for promoting neurodegeneration are *Drosophila* and mouse model systems in which overexpression of tau results in accumulation of phosphorylated tau and causes neurodegeneration in the absence of tangles (Ishihara et al., 1999; Spittaels et al., 1999; Duff et al., 2000; Probst et al., 2000; Wittmann et al., 2001). In contrast to overexpression models, a loss of tau is observed in some forms of human dementia that lack distinctive histopathology (Zhukareva et al., 2001). Thus, either overabundance or underabundance of tau disrupts the normal balance of tau functions and results in neurodegeneration without the requisite formation of NFTs. On the other hand, additional toxicities may stem directly from NFTs (see below).

The fact that Aβ can cause hyperphosphorylation of tau, the formation of dystrophic neurites, and cell death in cultured neurons is consistent with the activation of cellular mechanisms that are representative of the degenerative process in AD brain. This is in spite of the fact that neurons treated with fibrillar Aβ do not form NFTs. Of course, it may always be argued that neuronal cultures from animals that do not normally form NFTs do not adequately model the human disease condition. This simple cellular model neither negates, nor even addresses, a specific role for NFTs in the human disease process.

Attempts have been made to recreate NFTs *in vitro* and in cellular model systems and to use these systems to study possible mechanisms by which NFTs form *in vivo*. *In vitro* studies suggest that tau aggregation into filamentous forms is promoted by polyanions, such as sulfated glycosaminoglycans (Goedert et al., 1996; Perez et al., 1996) and RNA (Kampers et al., 1996), or by oxidized fatty acids (Gamblin et al., 2000). The formation of tau fibrils has also been demonstrated in cellular model systems. Exposure of cultured rat hippocampal and human cortical neurons to the lipid peroxidation product 4-hydroxynonenal caused hyperphosphorylation by impairing the ability to phosphatases to dephosphorylate tau (Mattson et al., 1997). SH-SY5Y cells treated with a combination of the phosphatase inhibitor, okadaic acid, and 4-hydroxynonenal, formed filamentous aggregates of phosphorylated tau (Perez et al., 2002). The fact that inhibition of phosphatase activity promotes the formation of tau filaments in a cellular model is consistent with the concept that increased tau phosphorylation precedes and promotes NFT formation in AD brain. Recent animal models also support the role of tau phosphorylation in the formation of NFTs.

2. Recapitulating tau pathologies in animal models

The progression of tau pathology in AD brain can be divided into six stages according to the system introduced by Braak and Braak (1995). This staging system describes the progression of NFTs from the transentorhinal cortex (Stage I) ultimately to the neocortex (Stage VI). AD patients at Stage I are asymptomatic while those at Stage VI have the most severe form of the disease. At the cellular level neurons undergo a series of cytoskeletal changes on their way to demise (Braak et al., 1994). Tau, which normally resides in the axon, accumulates in the somatodendritic compartment in a hyperphosphorylated state. In this "pretangle" stage, neuronal morphology is normal and there is no sign of tau filament formation. In later stages, neurons develop intracellular NFTs consisting of PHFs and SFs, lose their processes and eventually die, leaving extracellular NFTs as "ghost tangles".

A number of mouse, rat, lamprey, and *Drosophila* model systems have been utilized in an attempt to recapitulate the above tau pathologies in AD. While there is still no widely accepted animal model that mimics all aspects of the human AD condition, each of these models addresses different aspects of the pathologies and functional changes associated with development of the disease, including increased tau phosphorylation, somatodendritic relocalization of tau, NFT deposition, and cell loss.

2.1. *Environmental stimuli modify tau phosphorylation*

One of the many hypotheses invoked to explain the age-related development of AD is the idea that chronic environmental stressors produce biochemical changes that initiate or promote the disease process. For example, AD patients display an exaggerated cortisol response to dexamethasone challenge (Greenwald, 1986; De Leon 1988), suggesting altered function of their hypothalamic–pituitary–adrenal axis. Several studies have addressed the effects of environmental stressors on tau phosphorylation *in vivo*. Among the earliest of these studies was the demonstration by Papasozomenos and Su (1991) that exposure of female rats to heat shock resulted in a delayed increase in tau phosphorylation 6 h later. A separate study by Korneyev et al. (1995) showed an immediate increase in tau phosphorylation following cold-water stress. Comparing these two models, heat shock produced a similar increase in tau phosphorylation as did cold-water stress, but the time courses of the effects were very different. Subsequent studies by Papasozomenos and Su (1995) in cerebral explants appeared to identify an opposing result, namely a rapid dephosphorylation of tau in response to heat shock. This rapid dephosphorylation of tau immediately after heat shock was subsequently observed *in vivo*, revealing a biphasic response to this stimulus (immediate dephosphorylation followed by delayed hyperphosphorylation) (Papasozomenos, 1996). Whether increases or decreases in tau phosphorylation are specifically responses to stress, or are merely responses to changes in body temperature, is unclear; however, the response to cold-water "stress" was unaffected by adrenalectomy (Korneyev et al., 1995), indicating that the hypothalamic–pituitary–adrenal axis is not involved in this response. Additional "stressors" reported to increase tau phosphorylation in animal models include ischemia (Geddes et al., 1994; Dewar and Dawson, 1995), seizures (Stein-Behrens et al., 1994), and fasting (Yanagisawa et al., 1999; Planel et al., 2001). Interestingly, fasted mice also display reduced body temperature (Dubuc et al., 1985), making body-temperature changes a common feature among many of these models of environmentally induced changes in tau phosphorylation.

Mechanisms underlying changes in tau phosphorylation have been investigated in the heat shock and fasting models. Increases in activities of three tau kinases, cdk5, GSK3β, and JNK, were reported in the heat-shock model (Shanavas and Papasozomenos, 2000). The increase in GSK3β activity, as well as the increase in tau phosphorylation, was blocked by pretreatment of female rats with testosterone propionate (Papasozomenos and Shanavas, 2002), consistent with a participation of GSK3β in the regulation of tau phosphorylation in this paradigm. In contrast, GSK3β and cdk5 activities were reduced in the fasting model (Planel et al., 2001). However, tau phosphatase activity, predominantly comprised of PP-2A, was also markedly reduced during fasting, thereby more than compensating for the reduction of tau kinase activities.

The animal studies described above highlight the fact that the phosphorylation status of tau is part of a dynamic adaptive response to changing environmental conditions. The recognition that tau kinase and phosphatase activities are susceptible to regulation by environmental stimuli may bear relevance to mechanisms underlying increased tau phosphorylation in AD, since PP-2A and PP-1 activities are reduced by 20–30% in AD brain (Gong et al., 1993, 1995) and the activities of a number of tau kinases are reported to be increased (Ghoshal et al., 1999; Hensley et al., 1999; Patrick et al., 1999; Pei et al., 1999; Perry et al., 1999; Yasojima et al., 2000; Zhu et al., 2000). The mechanisms underlying these changes have not been completely addressed. In the case of cdk5, increased proteolytic activity of calpain in AD brains (Grynspan et al., 1997) may lead to higher levels of p25 (Patrick et al., 1999; Tseng et al., 2002), which is more stable (Patrick et al., 1999) and has higher activity than its precursor p35 in stimulating cdk5-mediated tau phosphorylation (Hashiguchi et al., 2002). Another hypothesis is that aging impairs the responsiveness of the normal adaptive mechanisms, resulting in imbalance of kinase and phosphatase activities. This is exemplified by a report that GSK3β activity steadily increases in adult human brain beginning at about 60 years (Planel et al., 2002).

2.2. The pretangle stage can be mimicked by altered expression of tau phosphatases or tau kinases

Bancher et al. (1989) found that abnormal tau phosphorylation was not only found in tangle-bearing neurons but also in the cytoplasm of neurons that do not harbor NFTs, suggesting that somatodendritic accumulation of phosphorylated tau protein in the pretangle stage may be an early step in the development of tangles. Two groups of animal models mimic aspects of the pretangle stage. One group models the pretangle stage by induction of tau phosphorylation, while another group achieves somatodendritic localization through tau overexpression (Fig. 1). Animal models have been engineered to increase tau phosphorylation by reduced expression of tau phosphatases or by increased expression of tau kinases. Attempts to produce mice with targeted knockout of tau phosphatase activities have met modest success. Knockout of the catalytic subunit Aα of PP-2B gave rise to increased tau phosphorylation that was limited to the mossy fibers of the hippocampus (Kayyali et al., 1997). Similar efforts to remove PP-2A activity were unsuccessful due to embryonic lethality (Götz et al., 1998). Kins et al. (2001) reported successful generation of a mouse with reduced PP-2A activity via expression of a dominant negative form of PP-2A. This mouse displayed tau hyperphosphorylation in cortex, hippocampus, and cerebellum, and accumulation of phosphorylated tau in somal and dendritic compartments, similar to the abnormal redistribution of tau seen in AD brain. However, no NFTs were found.

Transgenic mice overexpressing tau protein kinase activities also display somatodendritic localization of hyperphosphorylated tau. For instance, overexpression of p25, an activator of cdk5, in transgenic mice induced tau hyperphosphorylation in neuronal cell bodies (Ahlijanian et al., 2000; Bian et al.,

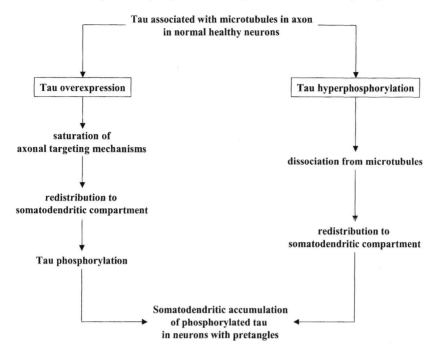

Fig. 1. Two potential mechanisms leading to the pretangle stage in neurons in transgenic mice. The pretangle stage is characterized by somatodendritic accumulation of hyperphosphorylated tau. Tau overexpression in transgenic mice saturates the axonal targeting mechanisms and redistrubutes tau to the somatodendritic compartment where it is phosphorylated by tau kinases (left side of flow chart). Alternatively, increasing tau phosphorylation by altering the kinase/phosphatase balance in transgenic animals dissociates tau from microtubules. Hyperphosphorylated tau then redistributes to the somatodendritic compartment, giving rise to neurons with pretangles (right side of flow chart).

2002). Three laboratories have reported increased tau phosphorylation following overexpression of GSK3β in mice (Brownlees et al., 1997; Spittaels et al., 2000; Lucas et al., 2001). In the models where immunocytochemistry was assessed, somatodendritic localization of phosphorylated tau was demonstrated. Similar to mice with diminished protein phosphatase activities, no NFTs were found in these kinase overexpressors. These findings suggest that tau phosphorylation is sufficient to induce the pretangle stage in neurons but insufficient for NFT formation.

2.3. The pretangle stage can also be induced by overexpression of tau proteins

Somatodendritic localization of phosphorylated tau has also been demonstrated in a number of animal models overexpressing human tau (Götz et al., 1995; Brion et al., 1999; Ishihara et al., 1999; Spittaels et al., 1999; Duff et al., 2000; Probst et al., 2000). It is not yet clear how overexpression of tau in transgenic mice increases tau phosphorylation and extends tau from its normal location in the axon to the somatodendritic compartment. The overall levels of tau expression in transgenic mice can be quite variable: from 0.1-fold of endogenous tau level (Götz et al., 1995) to 18-fold (Tatebayashi et al., 2002). However, the levels of tau expression in affected neurons could actually be much higher than the overall levels since human tau may be expressed in only a small percentage of neurons (Götz et al., 1995). Therefore, it is conceivable that the somatodendritic localization of overexpressed tau could be due to saturation of the mechanisms targeting tau to the axon, leading to an overflow of tau into the soma and dendrites. Overexpression of tau may not change the stoichiometry of tau phosphorylation, but simply enhances the detection of phosphotau epitopes that normally exist below detectable levels in the somatodendritic compartment.

On the other hand, there is some evidence that tau might be phosphorylated after its translocation to the somatodendritic compartment. While AT8 and PHF-1 staining were observed in the cell body, axons, and dendrites of some neurons in the transgenic mice expressing the longest human tau isoform, Götz et al. (1995) found these to be only a subset of the neurons with somatodendritic localization of tau. GSK-3, found in the cell body in wild-type mice and increased in the apical dendrites in tau transgenic mice (Götz and Nitsch, 2001), could be responsible for tau phosphorylation in the somatodendritic compartment (Fig. 1).

2.4. Formation of NFTs in animal models

In spite of the apparent importance of tau phosphorylation in the enhancement of NFT formation, the models described above demonstrate that increased tau phosphorylation, by itself, does not appear to be sufficient to initiate this process. It is unclear what other factors or conditions are required, in addition to tau phosphorylation, to form NFTs *in vivo*. Unlike humans, aged mice do not normally develop NFTs. Nelson et al. (1996) suggested that differences between the primary sequences of human and mouse tau might account for a greater propensity of human tau to form NFTs. To our knowledge, all transgenic animal models that exhibit NFTs overexpress human tau proteins. None of the models based on endogenous mouse tau, e.g., transgenics of tau kinases or phosphatase knockout, display NFTs. While these observations seem to be consistent with the view that human tau may have a property that makes it susceptible to the induction of NFT formation, the results could simply be due to the higher interest in human tau than in mouse tau and to the disproportionate number of models pursued using the human tau constructs. *In vitro* studies show that mouse tau aggregates into PHF with similar efficiency as that of human tau (Kampers et al., 1999), suggesting that mouse tau has similar intrinsic ability to form NFTs. Another view is that mice simply do not live long enough to develop NFTs. However, transgenic mouse models described below clearly demonstrate that it is possible to develop NFTs within the life span of a mouse.

2.4.1. Formation of NFTs can be induced by overexpression of tau proteins

Although many transgenic mice overexpressing wild-type tau reach the pretangle stage with somatodendritic accumulation of phosphorylated tau, most of these models do not develop NFTs. One exception, a transgenic mouse expressing the shortest isoform of wild-type tau, did appear to develop thioflavin-S positive and Congo Red-stained NFT-like inclusions, consisting mainly of SFs similar to those found in AD (Ishihara et al., 2001). In lampreys overexpressing the shortest human tau isoform, straight tau filaments, though thioflavin-S and Gallyas silver negative, were also present (Hall et al., 1997). In contrast to the majority of transgenic models expressing wild-type tau that do not form tangles, most reports of transgenic mice expressing human tau with FTDP-17 mutations indicate that these mice do go on to develop NFTs (Lewis et al., 2000; Götz et al., 2001a; Miyasaka et al., 2001; Tanemura et al., 2001; Allen et al., 2002; Tatebayashi et al., 2002). Therefore, it appears that FTDP-17 mutant tau proteins might induce tangle formation more readily than their wild-type counterparts, consistent with their autosomal dominant trait of inheritance (Wilhelmsen et al., 1994). *In vitro* studies show that some of the FTDP-17 mutant tau proteins have an increased propensity to aggregate (Barghorn et al., 2000) and reduced ability to bind to microtubules (Hong et al., 1998), which in turn may further enhance tau aggregation by increasing the amount of tau available (Nagiec et al., 2001). Thus, in contrast to wild-type tau, the FTDP-17 tau mutants may fulfill specific conditions required for tau aggregation in these transgenic animals.

In humans with frontotemporal dementia, some of the FTDP-17 mutations increase the ratio of 4 repeat to 3 repeat tau, suggesting that changing the ratio of 4 to 3 repeat tau may predispose an individual to developing tau pathologies (see review in Ingram and Spillantini, 2002). However, there does not appear to be a clear relationship between expression of specific tau isoforms and the ability to form tangles in transgenic mice. For instance, overexpression of wild-type tau containing either 3 repeats (Brion et al., 1999) or 4 repeats (Götz et al., 1995; Spittaels et al., 1999; Duff et al., 2000; Probst et al., 2000) has resulted only in neurons displaying a pretangle-like stage of somatodendritic phosphorylated tau localization. On the other hand, NFTs have been found in mice overexpressing the shortest 3 repeat wild-type tau isoform (Ishihara et al., 2001), or FTDP-17 mutant tau comprised of either the shortest 4 repeat isoform (Lewis et al., 2000; Allen et al., 2002) or the longest 4 repeat tau isoform (Götz and Nitsch, 2001; Tanemura et al., 2001; Tatebayashi et al., 2002).

In AD about 95% of the tau filaments have the paired helical structure while about 5% are SFs. Patients with specific FTDP-17 mutations also carry characteristic filamentous tau structures (see review in Ingram and Spillantini, 2002). For example, patients with V337M or R406W carry both PHF and SF while others have only twisted (P301L and K369I) or only straight tau filaments (R5H, R5L, and S320F). It is possible that the small changes in the primary sequence of tau may determine the structure of tau filaments. However, the structures of tau filaments in tau transgenic mice do not exactly correspond to those found in human cases carrying the same mutation. SFs are found in transgenic mice overexpressing the wild type (Ishihara et al., 2001), V337 (Tanemura et al., 2002), and R406W (Tatebayashi et al., 2002) FTDP-17 mutant tau proteins. Half-twisted filaments are found in the P301S transgenic (Allen et al., 2002). Interestingly, SFs were found in the P301L mice by Lewis et al. (2000) while both straight and twisted filaments were reported in another transgenic mouse carrying the same mutation (Götz et al., 2001a). A *Drosophila* model, in which the longest form of wild-type human tau was coexpressed with the GSK3β homolog *shaggy*, is the only transgenic model reported to date to contain both PHFs and SFs, similar to those in AD (Jackson et al., 2002).

2.4.2. Factors regulating NFT formation in animal models

Factors shown to regulate development of NFTs *in vivo* in genetically engineered animals include amyloid peptides, phosphorylation, and aging.

2.4.2.1. Amyloid peptides

According to the amyloid cascade hypothesis, pathologies, neuronal dysfunction, and death ultimately stem from the accumulation of amyloid peptides (Hardy and Selkoe, 2002). This hypothesis implies that amyloid peptides should be able to induce NFT formation. In 2001, two independent studies provided evidence that amyloid peptides can indeed enhance NFT formation *in vivo*. Lewis et al. crossed an APP transgenic mice, Tg2576, with the P301L mutant tau mice (Lewis et al., 2000) and observed an increased density of NFTs in the limbic region and olfactory cortex (Lewis et al., 2001). Similarly, when P301L mice (Götz et al., 2001a) were injected with Aβ42 fibrils, increased NFT deposits were seen (Götz et al., 2001b). The site of NFT deposition was found at some distance from the site of injection, corresponding to the location of the cell bodies from neurons that project to the injection site. This suggests that Aβ42 fibrils are able to potentiate somatodendritic NFT formation by stimulating the corresponding nerve terminals. This spatial separation of NFTs from the sites where Aβ is injected is analogous to the spatial separation of the tangle and plaque pathologies found in AD patients (Price and Sisodia, 1994).

Although amyloid peptides can enhance NFT development, it appears that they are not sufficient to initiate tau filament formation *in vivo*. For example, Aβ fibril injection increased NFTs in P301L mice, but not in wild-type mice (Götz et al., 2001b). Also, despite abundant amyloid burden, APP single transgenics lack NFTs (Moechars et al., 1999). One explanation is that endogenous mouse tau proteins are unable to form filaments deposited in NFTs. However, Kampers et al. (1999) showed that mouse tau proteins were able to form PHF *in vitro* and that mouse tau forms filaments as easily as human tau. Both studies showing the ability of amyloid peptides to increase NFTs used the P301L mice that are capable of developing NFTs in the absence of Aβ fibrils. Therefore, we speculate that Aβ fibrils are only able to potentiate NFTs in development but are unable to initiate this process (Fig. 2).

The complete mechanism by which amyloid peptides enhance NFT formation is not known, but it is known that fibrillar Aβ stimulates tau phosphorylation in

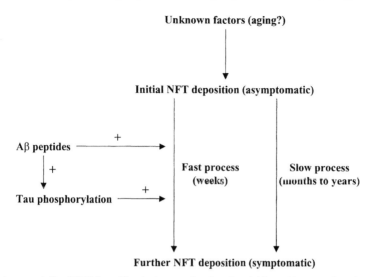

Fig. 2. Factors regulating NFT deposition in transgenic mice. Amyloid peptides, tau phosphorylation and aging have been implicated in the formation of NFTs in transgenic mice. Amyloid peptides and tau phosphorylation have been shown to potentiate NFT formation in transgenic mice in which NFTs are already present but not in those unable to form NFTs on their own. Therefore, it appears that amyloid peptides and tau phosphorylation can enhance NFT formation but are unable to initiate this process. Other unknown but essential factors are required to trigger NFT formation.

cultured neurons (Takashima et al., 1998; Alvarez et al., 2002; Zheng et al., 2002). APP overexpression in transgenic mice also induces tau hyperphosphorylation (Moechars et al., 1999; Tomidokoro et al., 2001). Given the facilitatory role of tau phosphorylation in NFT development, this may participate in the mechanism by which amyloid peptides affect NFT development.

2.4.2.2. Phosphorylation

To date two animal models demonstrate that protein phosphorylation, likely tau phosphorylation, promotes the formation of NFTs. The first model is in *Drosophila* overexpressing both human tau and the GSK3β homolog, *shaggy* (Jackson et al., 2002). NFTs stained positively by AT100 were strikingly increased in the double transgenic flies whereas only traces of diffuse staining were found in the flies expressing human tau alone. Both PHFs and SFs were found using electron microscopy. In a second model, cross-breeding of a p25 transgenic mouse with a P301L transgenic potentiated NFT deposition (Noble et al., 2003). Interestingly, cross-breeding of p25 with 8c mice (wild-type mice overexpressing all six wild-type human isoforms reported in Duff et al., 2000) did not produce NFTs (Karen Duff, personal communication). Since the 8c mice do not normally develop NFTs, whereas the P301L normally do, this suggests that phosphorylation, like amyloid peptides, can enhance NFT formation in progress but may be insufficient to initiate the process (Fig. 2).

2.4.2.3. Aging

Age is the biggest risk factor for the development of AD. Accordingly, neurofibrillary changes in AD may take decades to develop (Ohm et al., 1995). Age dependence of NFT development has also been observed in transgenic mice. This point is well illustrated in two elegant studies on the development of NFTs over time in a transgenic mouse overexpressing the shortest isoform of human tau (Ishihara et al., 1999, 2001). Although filamentous tau inclusions were present in young transgenic mice (< 12 months), these tau filaments lacked a number of key features characteristic of filaments found in AD brains. For instance, they were not stained by thioflavin-S, Congo Red, or Gallyas silver impregnation, all of which give positive results for authentic NFTs. These tau filaments were found in proximal axonal regions besides the cell body and dendrites. They also contained neurofilaments and microtubules (Ishihara et al., 1999). In contrast, as the animals aged (> 18 months), the tau filaments displayed features characteristically found in NFTs from AD brain, e.g., positive staining by thioflavin-S, Congo Red, and Gallyas silver method and absence of neurofilaments and microtubules (Ishihara et al., 2001).

Development of AD-like tau filaments may take months in the tau transgenic mice and decades in humans. However, tau filament assembly can be detected in an hour or less *in vitro* (Alonso et al., 2001; Perez et al., 2002), suggesting that these structures can develop quite rapidly given the right conditions. The long duration required for tau filament formation in tau transgenic mice and humans is not well understood. However, this long process can be considerably shortened by the presence of Aβ42 fibrils. Injection of Aβ42 fibrils into the amygdala of P301L mice gave rise to NFTs in as rapidly as 18 days. Therefore, there appear to be two different mechanisms in the formation of NFTs in transgenic mice: a slow process which takes months to years by tau overexpression and a facilitated process, induced by amyloid fibrils, which occurs in only a few weeks (Fig. 2).

2.5. Recapitulating neurodegeneration associated with tau pathologies

Hypotheses on the mechanisms of toxicities arising from tau-related pathologies fit into two main schools of thought: the loss of normal tau functions due to tau hyperphosphorylation, and the gain of toxic functions from, but not limited to, tau aggregation into PHFs (Fig. 3). One of the normal functions of tau is to stabilize microtubules. When hyperphosphorylated, tau loses its affinity for microtubules leading to microtubule disassembly, impairment of axonal transport, disruption of neuronal morphology, axonal degeneration, and finally neuronal cell death (Fig. 3). According to the loss of function school, these neurodegenerative processes can occur at the pretangle stage when tau is hyperphosphorylated and redistributed to the somatodendritic compartment. Aggregation of hyperphosphorylated tau proteins into NFTs further reduces the effective tau concentration for maintaining microtubule integrity. On the other hand, the gain of toxic function school suggests that toxicities can arise from the formation of filaments comprised of

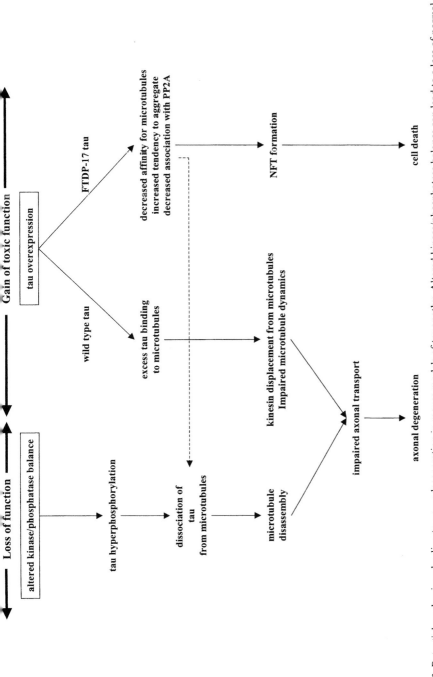

Fig. 3. Potential mechanisms leading to neurodegeneration in mouse models of tauopathy. Altered kinase/phosphatase balance can lead to a loss of normal tau functions. Tau hyperphosphorylation results in tau dissociation from microtubules, microtubule disassembly, impaired axonal transport, and degeneration. In contrast, tau overexpression studies suggest additional toxicities associated with a gain of toxic function for tau. Wild type tau overexpressed in transgenic mice competes for kinesin binding to microtubules and impairs microtubule dynamics, resulting in disruption of axonal transport and degeneration. Overexpression of FTDP-17 mutant tau gives rise to NFTs and may cause neuronal cell death. However, one transgenic mouse overexpressing wild type tau does appear to display NFTs (Ishihara et al., 2001)

hyperphosphorylated tau proteins (Fig. 3). Toxic functions may also arise from other properties of the abnormally phosphorylated tau in addition to NFT deposition.

Several animal models are consistent with the loss of tau function hypothesis in the pathogenesis of AD. Some transgenic animal models that develop pretangles but lack NFTs do show neuronal abnormalities. These include transgenic mice that have activated tau protein kinase activities. For instance, overexpression of the cdk5 activator p25 in transgenic mice induced tau hyperphosphorylation in neuronal cell bodies but did not cause NFT formation (Ahlijanian et al., 2000; Bian et al., 2002). Yet these transgenic mice developed dilated axons with accumulation of mitochondria and cell organelles, resembling impairment of microtubule function and impairment of axonal transport. Motor abnormalities resulting from axonal degeneration were also found. Presumably, p25/cdk5-induced tau hyperphosphorylation led to dissociation of tau from microtubules and subsequent microtubule disassembly. Transgenic animals with modified expression of GSK3β (Lucas et al., 2001), apolipoprotein E4 (Tesseur et al., 2000), Niemann Pick type C-1 (*NPC-1*) gene (Bu et al., 2002), or superoxide dismutase (Nguyen et al., 2001) also give rise to tau hyperphosphorylation, neuronal abnormalities, and behavioral impairment but no reported NFTs. Although these animal models are consistent with loss of the microtubule-binding function of tau, it is difficult to prove that this is the source of toxicity. For instance, cdk5 and GSK3β can phosphorylate many other substrates besides tau (reviewed in Cohen and Frame, 2001; Smith and Tsai, 2002). One cannot easily eliminate a contribution to neurodegeneration by phosphorylation of these other protein substrates.

On the other hand, a number of findings seem to argue against the loss of function hypothesis. Removing tau completely from mice by gene targeting did not appear to cause neurodegeneration, although microtubule stability in small-caliber axons was decreased (Harada et al., 1994). Acute quantitative removal of tau by injection of an anti-tau antibody into the soma of neurons before they extend axons did not have an apparent effect on microtubule integrity (Tint et al., 1998). However, these arguments do have their caveats. Compensatory mechanisms during and after development in the tau knockout may explain the lack of neurodegenerative features. Growing axons from sympathetic neurons were used in the anti-tau antibody experiments and might not represent events occurring in mature neurons in the central nervous system in patients suffering from neurodegenerative diseases.

Other evidence suggests that tau plays an active role in neurodegeneration and argues for the gain of toxic function hypothesis. Neurons cultured from tau knockout mice were protected from Aβ-induced cell death (Rapoport et al., 2002). Although several interpretations of this result are possible, the data are consistent with an interpretation whereby the toxicity of Aβ is related to changes in the normal functions of tau. The autosomal dominant nature of FTDP-17 suggests a gain of toxic functions from the tau mutants. Moreover, overexpression of either wild-type or FTDP-17 tau in transgenic animals can result in neurodegeneration and behavioral impairment. For instance, two transgenic mice overexpressing a 4 repeat human tau protein under the control of the murine thy-1 promoter generated by two independent laboratories each developed spinal cord axonal degeneration with accompanying motor impairment (Spittaels et al., 1999; Probst et al., 2000). In each case enlarged axons were found with accumulation of neurofilaments, mitochondria, and other organelles, resembling a functional axotomy and impairment of axonal transport. Overexpressed human tau might compete with kinesin for binding to microtubules and thereby disrupt axonal transport, leading to axonal degeneration (Ebneth et al., 1998). This hypothesis is supported by the reversal of phenotypes in a tau transgenic mouse through coexpression of a constitutively active form of GSK3β (Spittaels et al., 2000). Presumably, GSK3β rescued the tau transgenic mice by phosphorylating and dissociating excessive tau from binding to microtubules, thereby restoring normal microtubule function.

Although signs of axonal degeneration and behavioral deficits have been found in different animal models of tauopathy, loss of neurons (about 50%) has been reported only in the P301L (Lewis et al., 2000) and P301S mutant tau transgenic mice (Allen et al., 2002) in which NFTs were found. This suggests that additional toxicities might come from NFTs that push sick neurons over the edge to death. Exact mechanisms leading to neuronal cell death in these mice are unknown. Lysosomal abnormalities were observed in transgenic mice harboring three different FTDP-17 mutations of human tau and displaying tau filaments (Lim et al., 2001). *In vitro* studies show that proteasome activity can be inhibited by PHF isolated from AD brains or assembled from recombinant tau protein (Keck et al., 2003). Pathological significance of these findings is supported by reduction of proteasome activity in AD brains compared to controls (Keck et al., 2003). Proteasome is normally responsible for metabolizing ubiquitinated proteins. Inhibition of proteasome activity by NFTs in AD brains could lead to accumulation of toxic proteins that eventually cause cell death.

3. Implications, applications, and directions

To date a number of animal models have been established in which tau pathologies exist together with neuronal and behavioral dysfunctions. These animal models point toward at least three different types of possible mechanisms leading to neuronal degeneration and death (Fig. 3). First, overexpression of tau may saturate targeting mechanisms in the axon and result in somatodendritic accumulation. Protein kinases in the cell body may subsequently phosphorylate tau leading to the pretangle stage. High levels of tau in the axon may also displace kinesin from microtubules, disrupt microtubule dynamics, impair axonal transport, and thereby lead to axonal degeneration. Alternatively, cytoskeletal abnormalities may be induced either by increases in protein kinase activities or by reduction of protein phosphatase activities, giving rise to neurons with somatodendritic accumulation of hyperphosphorylated tau. Presumably, dissociation of hyperphosphorylated tau from microtubules and subsequent microtubule destabilization lead to the degenerative processes. Nevertheless, damages caused by other phosphorylated proteins cannot be excluded. Finally, there are animal models in which NFTs are found. Among the three groups of animal models, the one with NFTs is the only group with reported neuronal cell death in addition to cytoskeletal abnormalities and axonal degeneration, suggesting that NFTs may cause additional toxicities to neurons.

The current animal models of tauopathy also provide valuable insights into the interaction between tau and amyloid peptides. According to the amyloid cascade hypothesis, Aβ peptides are the primary stimuli for development of AD-related pathologies. However, studies in several animal models suggest that Aβ peptides cannot initiate NFT formation, although they can potentiate this process, possibly via induction of tau phosphorylation (Fig. 2). These studies imply that NFTs need to be present before amyloid peptides can further exacerbate this pathology. Indeed, in the earliest stage of AD, NFTs are found in the transentorhinal region in the absence of any signs of amyloid deposits. It is speculated that amyloid peptides exacerbate and propagate this initial tau pathology and accelerate the progression of the disease. This is consistent with the fact that FAD mutations on APP and presenilins increase amyloid peptide production and also trigger early-onset AD.

In addition to contributing to our understanding of tau pathologies, these animal models may be used as preclinical models for testing the potential efficacy of new therapeutic agents. The range of animal systems modeling various aspects of tau pathologies provides a broad base for testing therapeutic agents with different mechanisms. For instance, tau aggregation inhibitors can be tested in animal models with NFTs while kinase inhibitors can be tested in those with increased tau phosphorylation.

As discussed above, much progress has been achieved recently through the numerous animal models that recapitulate formation of pretangles and of NFTs. However, other important aspects of the disease are still missing in these models. For instance, with the exception of the *Drosophila* model, tau filaments formed in different mouse models are mostly SFs and sometimes twisted. This structure is different from the paired helical structure seen in the majority of tau filaments in AD. With the exception of the R406W transgenic mice (Tatebayashi et al., 2002), the behavioral deficits in the presence of pretangles or NFTs often involve motor dysfunction, making assessment of cognitive impairment difficult. This may have to do with the promoters used with the transgene and regional expression of tau. To make regional expression of pathologies even more challenging is the temporal characteristics of NFT development in AD patients. NFTs first develop in the transentorhinal cortex and then propagate into the limbic region and finally the neocortex. Animal models with similar regional and temporal propagation of tau pathologies would be most interesting. Since the majority of AD cases are considered to be sporadic, environmental factors seem to play an important role in the pathogenesis of AD. Induction of tau pathologies in animal models by environmental stimuli may shed light on the prevention and treatment of AD.

References

Ahlijanian, M.K., Barrezueta, N.X., Williams, R.D., Jakowski, A., Kowsz, K.P., McCarthy, S., Coskran, T., Carlo, A., Seymour, P.A., Burkhardt, J.E., Nelson, R.B., McNeish, J.D., 2000. Hyperphosphorylated tau and neurofilament and cytoskeletal disruptions in mice overexpressing human p25, an activator of cdk5. Proc. Natl. Acad. Sci. 97, 2910–2915.

Albers, D.S., Augood, S.J., 2001. New insights into progressive supranuclear palsy. Trends Neurosci. 24, 347–352.

Allen, B., Ingram, E., Takao, M., Smith, M.J., Jakes, R., Virdee, K., Yoshida, H., Holzer, M., Craxton, M., Emson, P.C., Atzori, C., Migheli, A., Crowther, R.A., Ghetti, B., Spillantini, M.G., Goedert, M., 2002. Abundant tau filaments and nonapoptotic neurodegeneration in transgenic mice expressing human P301S tau protein. J. Neurosci. 22, 9340–9351.

Alonso, A.C., Grundke-Iqbal, I., Iqbal, K., 1996. Alzheimer's disease hyperphosphorylated tau sequesters normal tau into tangles of filaments and disassembles microtubules. Nature Med. 2, 783–787.

Alonso, A.D., Grundke Iqbal, I., Daria, H.S., Iqbal, K., 1997. Abnormal phosphorylation of tau and the mechanism of Alzheimer neurofibrillary degeneration: sequestration of microtubule-associated proteins 1 and 2 and the disassembly of microtubules by the abnormal tau. Proc. Natl. Acad. Sci. 94, 298–303.

Alonso, A.C., Zaidi, T., Grundke-Iqbal, I., Iqbal, K., 1994. Role of abnormally phosphorylated tau in the breakdown of microtubules in Alzheimer disease. Proc. Natl. Acad. Sci. USA 91, 5562–5566.

Alonso, A.D.C., Zaidi, T., Novak, M., Grundke-Iqbal, I., Iqbal, K., 2001. Hyperphosphorylation induces self-assembly of tau into tangles of paired helical filaments/straight filaments. Proc. Natl. Acad. Sci. 6923–6928.

Alvarez, G., Munoz-Montano, J.R., Satrustegui, J., Avila, J., Bogonez, E., Diaz-Nido, J., 2002. Regulation of tau phosphorylation and protection against beta-amyloid-induced neurodegeneration by lithium. Possible implications for Alzheimer's disease. Bipolar. Disord. 4, 153–165.

Arriagada, P.V., Marzloff, K., Hyman, B.T., 1992. Distribution of Alzheimer-type pathologic changes in nondemented elderly individuals matches the pattern in Alzheimer's disease. Neurology 42, 1681–1688.

Bancher, C., Brunner, C., Lassmann, H., Budka, H., Jellinger, K., Wiche, G., Seitelberger, F., Grundke-Iqbal, I., Iqbal, K., Wisniewski, H.M., 1989. Accumulation of abnormally phosphorylated tau precedes the formation of neurofibrillary tangles and neuropil threads. Brain Res. 477, 90–99.

Barghorn, S., Zheng-Fischhofer, Q., Ackmann, M., Biernat, J., von Bergen, M., Mandelkow, E.M., Mandelkow, E., 2000. Structure, microtubule interactions, and paired helical filament aggregation by tau mutants of frontotemporal dementias. Biochemistry 39, 11714–11721.

Berg, L., McKeel, D.W., Miller, J.P., Storandt, M., Rubin, E.H., Morris, J.C., Baty, J., Coats, M., Norton, J., Goate, A.M., Price, J.L., Gearing, M., Mirra, S.S., Saunders, A.M., 1998. Clinicopathologic studies in cognitively healthy aging and alzheimer-disease-relation of histologic markers to dementia severity, age, sex, and apolipoprotein e genotype. Arch. Neurol. 55, 326–335.

Bian, F., Nath, R., Sobocinski, G., Booher, R.N., Lipinski, W.J., Callahan, M.J., Pack, A., Wang, K.K., Walker, L.C., 2002. Axonopathy, tau abnormalities, and dyskinesia, but no neurofibrillary tangles in p25-transgenic mice. J. Comp. Neurol. 446, 257–266.

Bosanquet, N., 2001. The socioeconomic impact of Alzheimer's disease. Int. J. Ger. Psychiatry 16, 249–253.

Braak, H., Braak, E., 1995. Staging of Alzheimer's disease-related neurofibrillary changes. Neurobiol. Aging 16, 271–278 discussion 278–284.

Braak, H., Braak, E., 1998. Evolution of neuronal changes in the course of Alzheimer's disease. J. Neural. Transm. Suppl. 53, 127–140.

Braak, E., Braak, H., Mandelkow, E.-M., 1994. A sequence of cytoskeleton changes related to the formation of neurofibrillary tangles and neuropil threads. Acta Neuropathol. 87, 554–567.

Brion, J.-P., Tremp, G., Octave, J.-N., 1999. Transgenic expression of the shortest human tau affects its compartmentalization and its phosphorylation as in the pretangle stage of Alzheimer's disease. Am. J. Pathol. 154, 255–270.

Brownlees, J., Irving, N.G., Brion, J.P., Gibb, B.J.M., Wagner, U., Woodgett, J., Miller, C.C.J., 1997. Tau phosphorylation in transgenic mice expressing glycogen synthase kinase-3-beta transgenes. Neuroreport 8, 3251–3255.

Bu, B.T., Li, J., Davies, P., Vincent, I., 2002. Deregulation of cdk5, hyperphosphorylation, and cytoskeletal pathology in the Niemann-Pick type C murine model. J. Neurosci. 22, 6515–6525.

Buée, L., Bussière, T., Buée-Scherrer, V., Delacourte, A., Hof, P.R., 2000. Tau protein isoforms, phosphorylation and role in neurodegenerative disorders. Brain Res. Rev. 33, 95–130.

Cai, X.D., Golde, T.E., Younkin, S.G., 1993. Release of excess amyloid beta protein from a mutant amyloid beta protein precursor. Science 259, 514–516.

Citron, M., Oltersdorf, T., Haass, C., McConlogue, L., Hung, A.Y., Seubert, P., Vigo-Pelfrey, C., Lieburg, I., Selkoe, D.J., 1992. Mutation of the beta-amyloid precursor protein in familial Alzheimer's disease increases beta-protein production. Nature 360, 672–674.

Cohen, P., Frame, S., 2001. The renaissance of GSK3. Nat. Rev. Mol. Cell. Biol. 2, 769–776.

de Leon, M.J., McRae, T., Tsai, J.R., George, A.E., Marcus, D.L., Freedman, M., Wolf, A.P., McEwen, B., 1988. Abnormal cortisol response in Alzheimer's disease linked to hippocampal atrophy. Lancet 2, 391–392.

Dewar, D., Dawson, D., 1995. Tau protein is altered by focal cerebral ischaemia in the rat: an immunohistochemical and immunoblotting study. Brain Res. 684, 70–78.

Dubuc, P.U., Wilden, N.J., Carlisle, H.J., 1985. Fed and fasting thermoregulation in ob/ob mice. Ann. Nutr. Metab. 29, 358–365.

Duff, K., Knight, H., Refolo, L.M., Sanders, S., Yu, X., Picciano, M., Malester, B., Hutton, M., Adamson, J., Goedert, M., Burki, K., Davies, P., 2000. Characterization of pathology in transgenic mice over-expressing human genomic and cDNA tau transgenes. Neurobiol. Dis. 7, 87–98.

Ebneth, A., Godemann, R., Stamer, K., Illenberger, S., Trinczek, B., Mandelkow, E.M., Mandelkow, E., 1998. Overexpression of tau protein inhibits kinesin-dependent trafficking of vesicles, mitochondria, and endoplasmic reticulum – implications for Alzheimers-disease. J. Cell Biol. 143, 777–794.

Finch, C.E., Tanzi, R.E., 1997. Genetics of aging. Science 278, 407–411.

Gamblin, T.C., King, M.E., Kuret, J., Berry, R.W., Binder, L.I., 2000. Oxidative regulation of fatty acid-induced tau polymerization. Biochemistry 39, 14203–14210.

Geddes, J.W., Schwab, C., Craddock, S., Wilson, J.L., Pettigrew, L.C., 1994. Alterations in tau immunostaining in the rat hippocampus following transient cerebral ischemia. J. Cereb. Blood Flow Metab. 14, 554–564.

Ghoshal, N., Smiley, J.F., DeMaggio, A.J., Hoekstra, M.F., Cochran, E.J., Binder, L.I., Kuret, J., 1999. A new molecular link between the fibrillar and granulovacuolar lesions of Alzheimer's disease. Am. J. Pathol. 155, 1163–1172.

Giannakopoulos, P., Gold, G., Duc, M., Michel, J.P., Hof, P.R., Bouras, C., 2000. Neural substrates of spatial and temporal disorientation in Alzheimer's disease. Acta Neuropathol. 100, 189–195.

Goedert, M., Jakes, R., Spillantini, M.G., Hasegawa, M., Smith, M.J., Crowther, R.A., 1996. Assembly of microtubule-associated protein tau into Alzheimer-like filaments induced by sulphated glycosaminoglycans. Nature 383, 550–553.

Gong, C.-X., Singh, T.J., Grundke-Iqbal, I., Iqbal, K., 1993. Phosphoprotein phosphatase activities in Alzheimer disease brain. J. Neurochem. 61, 921–927.

Gong, C.-X., Shaikh, S., Wang, J.-Z., Zaidi, T., Grundke-Iqbal, I., Iqbal, K., 1995. Phosphatase activity toward abnormally phosphorylated tau: decrease in Alzheimer disease brain. J. Neurochem. 65, 732–738.

Götz, J., Chen, F., Barmettler, R., Nitsch, R.M., 2001a. Tau filament formation in transgenic mice expressing P301L tau. J. Biol. Chem. 276, 529–534.

Götz, J., Chen, F., van Dorpe, J., Nitsch, R.M., 2001b. Formation of neurofibrillary tangles in P301L tau transgenic mice induced by Aβ42 fibrils. Science 293, 1491–1495.

Götz, J., Nitsch, R.M., 2001. Compartmentalized tau hyperphosphorylation and increased levels of kinases in transgenic mice. NeuroReport 12, 2007–2016.

Götz, J., Probst, A., Ehler, E., Hemmings, B., Kues, W., 1998. Delayed embryonic lethality in mice lacking protein phosphatase 2a catalytic subunit c-alpha. Proc. Natl. Acad. Sci. USA 95, 12370–12375.

Götz, J., Probst, A., Spillantini, M.G., Schafer, M.G., Jakes, R., Bürki, K., Goedert, M., 1995. Somatodendritic localization and hyperphosphorylation of tau protein in transgenic mice expressing the longest human brain tau isoform. EMBO J. 14, 1304–1313.

Greenwald, B.S., Mathe, A.A., Mohs, R.C., Levy, M.I., Johns, C.A., Davis, K.L., 1986. Cortisol and Alzheimer's disease, II: Dexamethasone supression, dementia severity, and affective symptoms. Am. J. Psych. 143, 442–446.

Grynspan, F., Griffin, W.R., Cataldo, A., Katayama, S., Nixon, R.A., 1997. Active site directed antibodies identify calpain II as an early appearing and pervasive component of neurofibrillary pathology in Alzheimer's disease. Brain Res. 763, 145–158.

Hall, G.F., Yao, J., Lee, G., 1997. Human tau becomes phosphorylated and forms filamentous deposits when overexpressed in lamprey central neurons in situ. Proc. Natl. Acad. Sci. 94, 4733–4738.

Harada, A., Oguchi, K., Okabe, S., Kuno, J., Terada, S., Ohshima, T., Sato-Yoshitake, R., Takei, Y., Noda, T., Hirokawa, N., 1994. Altered microtubule organization in small-calibre axons of mice lacking tau protein. Nature 369, 488–491.

Hardy, J., Selkoe, D.J., 2002. Medicine – The amyloid hypothesis of Alzheimer's disease: progress and problems on the road to therapeutics. Science 297, 353–356.

Hashiguchi, M., Saito, T., Hisanaga, S., Hashiguchi, T., 2002. Truncation of CDK5 activator p35 induces intensive phosphorylation of Ser(202)/Thr(205) of human tau. J. Biol. Chem. 277, 44525–44530.

Hensley, K., Floyd, R.A., Zheng, N.Y., Nael, R., Robinson, K.A., Nguyen, X., Pye, Q.N., Stewart, C.A., Geddes, J., Markesbery, W.R., Johnson, G.V., Bing, G., 1999. p38 kinase is activated in the Alzheimer's disease brain. J. Neurochem. 72, 2053–2058.

Holcomb, L.A., Gordon, M.N., Jantzen, P., Hsiao, K., Duff, K., Morgan, D., 1999. Behavioral changes in transgenic mice expressing both amyloid precursor protein and presenilin-1 mutations: lack of association with amyloid deposits. Behav. Genet. 29, 177–185.

Hong, M., Zhukareva, V., Vogelsberg-Ragaglia, V., Wszolek, Z., Reed, L., Miller, B.I., Geschwind, D.H., Bird, T.D., McKeel, D., Goate, A., Morris, J.C., Wilhelmsen, K.C., Schellenberg, G.D., Trojanowski, J.Q., Lee, V.M.Y., 1998. Mutation-specific functional impairments in distinct Tau isoforms of hereditary FTDP-17. Science 282, 1914–1917.

Hutton, M., Lendon, C.L., Rizzu, P., Baker, M., Froelich, S., Houlden, H., Pickering-Brown, S., Chakraverty, S., Isaacs, A., Grover, A., Hackett, J., Adamson, J., Lincoln, S., Dickson, D., Davies, P., Petersen, R.C., Stevens, M., de Graaff, E., Wauters, E., van Baren, J., Hillebrand, M., Joosse, M., Kwon, J.M., Nowotny, P.; Heutink, P., et al., 1998. Association of missense and 5'-splice-site mutations in tau with the inherited dementia FTDP-17. Nature 393, 702–705.

Ingram, E.M., Spillantini, M.G., 2002. Tau gene mutations: dissecting the pathogenesis of FTDP-17. Trends Mol. Med. 8, 555–562.

Iqbal, K., Zaidi, T., Bancher, C., Grundke-Iqbal, I., 1994. Alzheimer paired helical filaments. Restoration of the biological activity by dephosphorylation. FEBS Lett. 349, 104–108.

Ishihara, T., Hong, M., Zhang, B., Nakagawa, Y., Lee, M.K., Trojanowski, J.Q., Lee, V.M.-Y., 1999. Age-dependent emergence and progression of a tauopathy in transgenic mice overexpressing the shortest human tau isoform. Neuron 24, 751–762.

Ishihara, T., Zhang, B., Higuchi, M., Yoshiyama, Y., Trojanowski, J.Q., Lee, V.M.-Y., 2001. Age-dependent induction of congophilic neurofibrillary tau inclusions in tau transgenic mice. Am. J. Pathol. 158, 555–562.

Jackson, G.R., Wiedau-Pazos, M., Sang, T.K., Wagle, N., Brown, C.A., Massachi, S., Geschwind, D.H., 2002. Human wild-type tau interacts with wingless pathway components and produces neurofibrillary pathology in Drosophila. Neuron 34, 509–519.

Johnson, N., Davis, T., Bosanquet, N., 2000. The epidemic of Alzheimer's disease. How can we manage the costs? Pharmacoeconomics 3, 215–223.

Kampers, T., Friedhoff, P., Biernat, J., Mandelkow, E.M., Mandelkow, E., 1996. RNA stimulates aggregation of microtubule-associated protein tau into Alzheimer-like paired helical filaments. FEBS Lett. 399, 344–349.

Kampers, T., Pangalos, M., Geerts, H., Wiech, H., Mandelkow, E., 1999. Assembly of paired helical filaments from mouse tau: implications for the neurofibrillary pathology in transgenic mouse models for Alzheimer's disease. FEBS Lett. 451, 39–44.

Kawas, C.H., Katzman, R., 1999. Epidemiology of dementia and Alzheimer disease. In: R.D. Terry, R. Katzman, K.L. Bick, S.S. Sisodia (Eds.), Alzheimer Disease. Lippincott, Williams & Wilkins, Philadelphia, pp. 95–116.

Kayyali, U.S., Zhang, W., Yee, A.G., Seidman, J.G., Potter, H., 1997. Cytoskeletal changes in the brains of mice lacking calcineurin A alpha. J. Neurochem. 68, 1668–1678.

Keck, S., Nitsch, R., Grune, T., Ullrich, O., 2003. Proteasome inhibition by paired helical filament-tau in brains of patients with Alzheimer's disease. J. Neurochem. 85, 115–122.

Kins, S., Crameri, A., Evans, D.R.H., Hemmings, B.A., Nitsch, R.M., Götz, J., 2001. Reduced protein phosphatase 2A activity induces hyperphosphorylation and altered compartmentalization of tau in transgenic mice. J. Biol. Chem. 276, 38193–38200.

Korneyev, A., Binder, L., Bernardis, J., 1995. Rapid reversible phosphorylation of rat brain tau proteins in response to cold water stress. Neurosci. Lett. 191, 19–22.

Lee, V.M.-Y., 2001. Tauists and βaptists united-well almost! Science 293, 1146–1147.

Lee, V.M.-Y., Goedert, M., Trojanowski, J.Q., 2001. Neurodegenerative tauopathies. Ann. Rev. Neurosci. 24, 1121–1159.

Lewis, J., Dickson, D.W., Lin, W.-L., Chisholm, L., Corral, A., Jones, G., Yen, S.-H., Sahara, N., Skipper, L., Yager, D., Eckman, C., Hardy, J., Hutton, M., McGowan, E., 2001. Enhanced neuro-fibrillary degeneration in transgenic mice expressing mutant tau and APP. Science 293, 1487–1491.

Lewis, J., McGowan, E., Rockwood, J., Melrose, H., Nacharaju, P., Van Slegtenhorst, M., Gwinn-Hardy, K., Murphy, M.P., Baker, M., Yu, X., Duff, K., Hardy, J., Corral, A., Lin, W.L., Yen, S.H., Dickson, D.W., Davies, P., Hutton, M., 2000. Neurofibrillary tangles, amyotrophy and progressive motor disturbance in mice expressing mutant (P301L) tau protein. Nature Genet. 25, 402–405.

Lim, F., Hernandez, F., Lucas, J.J., Gomez-Ramos, P., Moran, M.A., Avila, J., 2001. FTDP-17 mutations in tau transgenic mice provoke lysosomal abnormalities and tau filaments in forebrain. Mol. Cell. Neurosci. 18, 702–714.

Lucas, J.J., Hernández, F., Gómez-Ramos, P., Morán, M.A., Hen, R., Avila, J., 2001. Decreased nuclear β-catenin, tau hyperphosphorylation and neurodegeneration in GSK-3β conditional transgenic mice. EMBO J. 20, 27–39.

Mattson, M.P., Fu, W., Waeg, G., Uchida, K., 1997. 4-Hydroxynonenal, a product of lipid peroxidation, inhibits dephosphorylation of the micortubule-associated protein tau. Neuroreport 8, 2275–2281.

Miyasaka, T., Morishima-Kawashima, M., Ravid, R., Heutink, P., van Swieten, J.C., Nagashima, K., Ihara, Y., 2001. Molecular analysis of mutant and wild-type tau deposited in the brain affected by the FTDP-17 R406W mutation. Am. J. Pathol. 158, 373–379.

Moechars, D., Dewachter, I., Lorent, K., Reverse, D., Baekelandt, V., Naidu, A., Tesseur, I., Spittaels, K., Van Den Haute, C., Checler, F., Godaux, E., Cordell, B., Van Leuven, F., 1999. Early phenotypic changes in transgenic mice that overexpress different mutants of amyloid precursor protein in brain. J. Biol. Chem. 274, 6483–6492.

Mudher, A., Lovestone, S., 2002. Alzheimer's disease – do tauists and baptists finally shake hands? Trends Neurosci. 25, 22–26.

Nagicc, E.E., Sampson, K.E., Abraham, I., 2001. Mutated tau binds less avidly to microtubules than wildtype tau in living cells. J. Neurosci. Res. 63, 268–275.

Nagy, Z., Esiri, M.M., Jobst, K.A., Morris, J.H., King, E.M., McDonald, B., Litchfield, S., Smith, A., Barnetson, L., Smith, A.D., 1995. Relative roles of plaques and tangles in the dementia of Alzheimer's disease: correlations using three sets of neuropathological criteria. Dementia 6, 21–31.

Nelson, P.T., Stefansson, K., Gulcher, J., Saper, C.B., 1996. Molecular evolution of tau protein – implications for Alzheimers disease. J. Neurochem. 67, 1622–1632.

Nguyen, M.D., Lariviere, R.C., Julien, J.-P., 2001. Deregulation of cdk5 in a mouse model of ALS: toxicity alleviated by perikaryal neurofilament inclusions. Neuron 30, 135–147.

Noble, W., Olm, V., Takata, K., Casey, E., Mary, O., Meyerson, J., Gaynor, K., LaFrancois, J., Wang, L., Kondo, T., Davies, P., Burns, M., Veeranna, Nixon, R., Dickson, D., Matsuoka, Y., Ahlijanian, M., Lau, L.F., Duff, K., 2003. Cdk5 is a key factor in tau aggregation and tangle formation in vivo. Neuron 38, 555–565.

Ohm, T.G., Muller, H., Braak, H., Bohl, J., 1995. Close-meshed prevalence rates of different stages as a tool to uncover the rate of Alzheimer's disease-related neurofibrillary changes. Neuroscience 64, 209–217.

Papasozomenos, S., Shanavas, A., 2002. Testosterone prevents the heat shock-induced overactivation of glycogen synthase kinase-3 beta but not of cyclin-dependent kinase 5 and c-Jun NH2-terminal kinase and concomitantly abolishes hyperphosphorylation of tau: implications for Alzheimer's disease. Proc. Natl. Acad. Sci. USA 99, 1140–1145.

Papasozomenos, S.C., 1996. Heat shock induces rapid dephosphorylation of tau in both female and male rats followed by hyperphosphorylation only in female rats: implications for Alzheimer's disease. J. Neurochem. 66, 1140–1149.

Papasozomenos, S.C., Su, Y., 1991. Altered phosphorylation of tau protein in heat-shocked rats and patients with Alzheimer disease. Proc. Natl. Acad. Sci. USA 88, 4543–4547.

Papasozomenos, S.C., Su, Y., 1995. Rapid dephosphorylation of tau in heat-shocked fetal rat cerebral explants: prevention and hyperphosphorylation by inhibitors of protein phosphatases PP1 and PP2A. J. Neurochem. 65, 396–406.

Patrick, G.N., Zukerberg, L., Mikolic, M., de la Monte, S., Dikkes, P., Tsai, L.-H., 1999. Conversion of p35 to p25 deregulates cdk5 activity and promotes neurodegeneration. Nature 402, 615–622.

Pei, J.-J., Braak, E., Braak, H., Grundke-Iqbal, I., Iqbal, K., Winblad, B., Cowburn, R.F., 1999. Distribution of active glycogen synthase kinase 3β (GSK3β) in brains staged for Alzheimer disease neurofibrillary changes. J. Neuropathol. Exp. Neurol. 58, 1010–1019.

Perez, M., Hernandez, F., Gomez-Ramos, A., Smith, M., Perry, G., Avila, J., 2002. Formation of aberrant phosphotau fibrillar polymers in neural cultured cells. Eur. J. Biochem. 269, 1484–1489.

Perez, M., Valpuesta, J.M., Medina, M., Montejo de Garcini, E., Avila, J., 1996. Polymerization of tau into filaments in the presence of heparin: the minimal sequence required for tau–tau interaction. J. Neurochem. 67, 1183–1190.

Perry, G., Roder, H., Nunomura, A., Takeda, A., Friedlich, A.L., Zhu, X., Raina, A.K., Holbrook, N., Siedlak, S.L., Harris, P.L., Smith, M.A., 1999. Activation of neuronal extracellular receptor kinase (ERK) in Alzheimer disease links oxidative stress to abnormal phosphorylation. NeuroReport 10, 2411–2415.

Planel, E., Yasutake, K., Fujita, S.C., Ishiguro, K., 2001. Inhibition of protein phosphatase 2A overrides tau protein kinase I/glycogen synthase kinase 3β and cyclin-dependent kinase 5 inhibition and results in tau hyperphosphorylation in the hippocampus of starved mouse. J. Biol. Chem. 276, 34298–34306.

Planel, E., Sun, X.Y., Takashima, A., 2002. Role of GSK-3β in Alzheimer's disease pathology. Drug Dev. Res. 56, 491–510.

Poorkaj, P., Bird, T.D., Wijsman, E., Nemens, E., Garruto, R.M., Anderson, L., Andreadis, A., Wiederholt, W.C., Raskind, M., Schellenberg, G.D., 1998. Tau is a candidate gene for chromosome 17 frontotemporal dementia. Ann. Neurol. 43, 815–825.

Price, D.L., Sisodia, S.S., 1994. Cellular and molecular biology of Alzheimer's disease and animal models. Ann. Rev. Med. 45, 435–446.

Probst, A., Götz, J., Wiederhold, K.H., Tolnay, M., Mistl, C., Jaton, A.L., Hong, M., Ishihara, T., Lee, V.M.-Y., Trojanowski, J.Q., Jakes, R., Crowther, R.A., Spillantini, M.G., Bürki, K., Goedert, M., 2000. Axonopathy and amyotrophy in mice transgenic for human four-repeat tau protein. Acta Neuropathol. 99, 469–481.

Rapoport, M., Dawson, H.N., Binder, L.I., Vitek, M.P., Ferreira, A., 2002. Tau is essential to beta-amyloid-induced neurotoxicity. Proc. Natl. Acad. Sci. USA 99, 6364–6369.

Selkoe, D., 2001. Clearing the brain's amyloid cobwebs. Neuron 32, 177–180.

Shanavas, A., Papasozomenos, S.C., 2000. Tau kinases in the rat heat shock model: possible implications for Alzheimer disease. Proc. Natl. Acad. Sci. 26, 14139–14144.

Smith, D.S., Tsai, L.-H., 2002. Cdk5 behind the wheel: a role in trafficking and transport? Trends Cell Biol. 12, 28–36.

Spillantini, M.G., Goedert, M., 1998. Tau protein pathology in neurodegenerative diseases. Trends Neurosci. 21, 428–433.

Spittaels, K., van den Haute, C., van Dorpe, J., Bruynseels, K., Vandezande, K., Laenen, I., Geerts, H., Mercken, M., Sciot, R., van Lommel, A., Loos, R., van Leuven, F., 1999. Prominent axonopathy in the brain and spinal cord of transgenic mice overexpressing four-repeat human tau protein. Am. J. Pathol. 155, 2153–2165.

Spittaels, K., van den Haute, C., van Dorpe, J., Geerts, H., Mercken, M., Bruynseels, K., Lasrado, R., Vandezande, K., Laenen, I., Boon, T., van Lint, J., Vandenheede, J., Moechars, D., Loos, R., van Leuven, F., 2000. Glycogen synthase kinase-3β phosphorylates protein tau and rescues the axonopathy in the central nervous system of human four-repeat tau transgenic mice. J. Biol. Chem. 275, 41340–41349.

Stein-Behrens, B., Mattson, M.P., Chang, I., Yeh, M., Sapolsky, R., 1994. Stress exacerbates neuron loss and cytoskeletal pathology in the hippocampus. J. Neurosci. 14, 5373–5380.

Suzuki, N., Cheung, T.T., Cai, X.D., Odaka, A., Otvos, L., Jr., Eckman, C., Golde, T.E., Younkin, S.G., 1994. An increased percentage of long amyloid beta protein secreted by familial amyloid beta protein precursor (beta APP717) mutants. Science 264, 1336–1340.

Takashima, A., Honda, T., Yasutake, K., Michel, G., Murayama, O., Murayama, M., Ishiguro, K., Yamaguchi, H., 1998. Activation of tau protein kinase I/glycogen synthase kinase-3beta by amyloid beta peptide (25–35) enhances phosphorylation of tau in hippocampal neurons. Neurosci. Res. 31, 317–323.

Tanemura, K., Akagi, T., Murayama, M., Kikuchi, N., Murayama, O., Hashikawa, T., Yoshiike, Y., Park, J.M., Matsuda, K., Nakao, S., Sun, X.Y., Sato, S., Yamaguchi, H., Takashima, A., 2001. Formation of filamentous tau aggregations in transgenic mice expressing V337M human tau. Neurobiol. Dis. 8, 1036–1045.

Tanemura, K., Murayama, M., Akagi, T., Hashikawa, T., Tominaga, T., Ichikawa, M., Yamaguchi, H., Takashima, A., 2002. Neurodegeneration with tau accumulation in a transgenic mouse expressing V337M human tau. J. Neurosci. 22, 133–141.

Tatebayashi, Y., Miyasaka, T., Chui, D.-H., Akagi, T., Mishima, K.-I., Iwasaki, K., Fujiwara, M., Tanemura, K., Ishiguro, K., Planel, E., Sato, S., Hashikawa, T., Takashima, A., 2002. Tau filament formation and associative memory deficit in aged mice expressing mutant (R406W) human tau. Proc. Natl. Acad. Sci. 99, 13896–13901.

Tesseur, I., van Dorpe, J., Spittaels, K., van den Haute, C., Moechars, D., van Leuven, F., 2000. Expression of human apolipoprotein E4 in neurons causes hyperphosphorylation of protein tau in the brains of transgenic mice. Am. J. Pathol. 156, 951–964.

Tint, I., Slaughter, T., Fischer, I., Black, M.M., 1998. Acute inactivation of tau has no effect on dynamics of microtubules in growing axons of cultured sympathetic neurons. J. Neurosci. 18, 8660–8673.

Tomidokoro, Y., Ishiguro, K., Harigaya, Y., Matsubara, E., Ikeda, M., Park, J.-M., Yasutake, K., Kawarabayashi, T., Okamoto, K., Shoji, M., 2001. Aβ amyloidosis induces the initial stage of tau accumulation in APP$_{Sw}$ mice. Neurosci. Lett. 299, 169–172.

Trojanowski, J.Q., Lee, V.M.-Y., 2000. "Fatal attractions" of proteins. A comprehensive hypothetical mechanism underlying Alzheimer's Disease and other neurodegenerative disorders. Ann. N.Y. Acad. Sci. 924, 62–67.

Tseng, H.C., Zhou, Y., Shen, Y., Tsai, L.H., 2002. A survey of Cdk5 activator p35 and p25 levels in Alzheimer's disease brains. FEBS Lett. 523, 58–62.

Wang, J.-Z., Gong, C.-X., Zaidi, T., Grundke-Iqbal, I., Iqbal, K., 1995. Dephosphorylation of Alzheimer paired helical filaments by protein phosphatase-2A and -2B. J. Biol. Chem. 270, 4854–4860.

Wilhelmsen, K.C., Lynch, T., Pavlou, E., Higgins, M., Nygaard, T.G., 1994. Localization of disinhibition-dementia-parkinsonism-amyotrophy complex to 17q21-22. Am. J. Human Genet. 55, 1159–1165.

Wittmann, C.W., Wszolek, M.F., Shulman, J.M., Salvaterra, P.M., Lewis, J., Hutton, M., Feany, M.B., 2001. Tauopathy in Drosophila: neurodegeneration without neurofibrillary tangles. Science 293, 711–714.

Yanagisawa, M., Planel, E., Ishiguro, K., Fujita, S.C., 1999. Starvation induces tau hyperphosphorylation in mouse brain: implications for Alzheimer's disease. FEBS Lett. 461, 329–333.

Yasojima, K., Kuret, J., DeMaggio, A.J., McGeer, E., McGeer, P.L., 2000. Casein kinase 1 delta mRNA is upregulated in Alzheimer disease brain. Brain Res. 865, 116–120.

Zheng, W.H., Bastianetto, S., Mennicken, F., Ma, W., Kar, S., 2002. Amyloid beta peptide induces tau phosphorylation and loss of cholinergic neurons in rat primary septal cultures. Neuroscience 115, 201–211.

Zhu, X., Rottkamp, C.A., Boux, H., Takeda, A., Perry, G., Smith, M.A., 2000. Activation of p38 kinase links tau phosphorylation, oxidative stress, and cell cycle-related events in Alzheimer disease. J. Neuropathol. Exp. Neurol. 59, 880–888.

Zhukareva, V., Vogelsberg-Ragaglia, V., Van Deerlin, V.M., Bruce, J., Shuck, T., Grossman, M., Clark, C.M., Arnold, S.E., Masliah, E., Galasko, D., Trojanowski, J.Q., Lee, V.M., 2001. Loss of brain tau defines novel sporadic and familial tauopathies with frontotemporal dementia. Ann. Neurol. 49, 165–175.

List of Contributors

Peter W. Atadja   Novartis Institute for Biomedical Research
                             One Health Plaza
                             East Hanover, NJ 07936, USA
                             Phone: 1-862-778-0435   Fax: 1-973-781-7578
                             Email: peter.atadja@pharma.novartis.com

Bulbul Chakravarti   Keck Graduate Institute of Applied Life Sciences
                             535 Watson Drive
                             Claremont, CA 91711, USA
                             Phone: 1-909-607-9525   Fax: 1-909-607-8086
                             Email: bulbul_chakravarti@kgi.edu

Deb N. Chakravarti   Keck Graduate Institute of Applied Life Sciences
                             535 Watson Drive
                             Claremont, CA 91711, USA

Maria Luiza A. Fernandes   Department of Pathology
                             State University of Rio de Janeiro
                             Brazil

Thomas C. Foster   McKnight Chair for Brain Research in Memory Loss
                             Evelyn F. & William L. McKnight Brain Institute
                             University of Florida College of Medicine
                             PO Box 100244
                             Gainesville, FL 32610-0244, USA
                             Phone: 1-352-392-4359   Fax: 1-352-392-8347
                             Email: foster@mbi.ufl.edu

Paul O. Kwon   Novartis Institute for Biomedical Research
                             One Health Plaza
                             East Hanover, NJ 07936, USA

Lit-Fui Lau   MS220-4013
                             CNS Discovery
                             Pfizer Global Research and Development
                             Groton, CT 06340, USA
                             Phone: 1-860-715-1921   Fax: 1-860-715-2349
                             Email: lit-fui_lau@groton.pfizer.com

Ching-Shwun Lin

Department of Urology
Knuppe Molecular Urology Laboratory
University of California at San Francisco
Box 1695, Mount Zion Campus
San Francisco, CA 94143-1695, USA
Phone: 1-415-353-7205   Fax: 1-415-353-9586
Email: clin@urol.ucsf.edu

Tom F. Lue

Department of Urology
Knuppe Molecular Urology Laboratory
University of California, San Francisco
CA 94143-1695, USA

Eduardo M. Rocha

Department of Internal Medicine
State University of Campinas, Brazil

Charanjit Sandhu

Program in Proteomics and Bioinformatics
Banting and Best Department of Medical Research
University of Toronto
Toronto, ON
Canada
Email: charanjit.sandhu@utoronto.ca
Correspondence address:
CH Best Institute, University of Toronto
112 College Street, Rm 402
Toronto, ON
Canada M5G 1L6

Joel B. Schachter

CNS Discovery
Pfizer Global Research and Development
Groton, CT 06340, USA

Lício A. Velloso

Department of Internal Medicine
University of Campinas, UNICAMP
Campinas
SP 13083-970, Brazil
Phone: 55-19-3788-8950   Fax: 55-19-3788-8950
Email: lavelloso@fcm.unicamp.br

Advances in
Cell Aging and Gerontology
Series Editor: Mark P. Mattson
URL: http://www.elsevier.nl/locate/series/acag

Aims and Scope:

Advances in Cell Aging and Gerontology (ACAG) is dedicated to providing timely review articles on prominent and emerging research in the area of molecular, cellular and organismal aspects of aging and age-related disease. The average human life expectancy continues to increase and, accordingly, the impact of the dysfunction and diseases associated with aging are becoming a major problem in our society. The field of aging research is rapidly becoming the niche of thousands of laboratories worldwide that encompass expertise ranging from genetics and evolution to molecular and cellular biology, biochemistry and behavior. ACAG consists of edited volumes that each critically review a major subject area within the realms of fundamental mechanisms of the aging process and age-related diseases such as cancer, cardiovascular disease, diabetes and neurodegenerative disorders. Particular emphasis is placed upon: the identification of new genes linked to the aging process and specific age-related diseases; the elucidation of cellular signal transduction pathways that promote or retard cellular aging; understanding the impact of diet and behavior on aging at the molecular and cellular levels; and the application of basic research to the development of lifespan extension and disease prevention strategies. ACAG will provide a valuable resource for scientists at all levels from graduate students to senior scientists and physicians.

Books Published:

1. P.S. Timiras, E.E. Bittar, *Some Aspects of the Aging Process*, 1996, 1-55938-631-2
2. M.P. Mattson, J.W. Geddes, *The Aging Brain*, 1997, 0-7623-0265-8
3. M.P. Mattson, *Genetic Aberrancies and Neurodegenerative Disorders*, 1999, 0-7623-0405-7
4. B.A. Gilchrest, V.A. Bohr, *The Role of DNA Damage and Repair in Cell Aging*, 2001, 0-444-50494-X
5. M.P. Mattson, S. Estus, V. Rangnekar, *Programmed Cell Death, Volume I*, 2001, 0-444-50493-1
6. M.P. Mattson, S. Estus, V. Rangnekar, *Programmed Cell Death, Volume II*, 2001, 0-444-50730-2
7. M.P. Mattson, *Interorganellar Signaling in Age-Related Disease*, 2001, 0-444-50495-8
8. M.P. Mattson, *Telomerase, Aging and Disease*, 2001, 0-444-50690-X
9. M.P. Mattson, *Stem Cells: A Cellular Fountain of Youth*, 2002, 0-444-50731-0
10. M.P. Mattson, *Calcium Homeostasis and Signaling in Aging*, 2002, 0-444-51135-0
11. T. Hagen, *Mechanisms of Cardiovascular Aging*, 2002, 0-444-51159-8
12. M.P. Mattson, *Membrane Lipid Signaling in Aging and Age-Related Disease*, 2003, 0-444-51297-7
13. G. Pawelec, *Basic Biology and Clinical Impact of Immunosenescence*, 2003, 0-444-51316-7
14. M.P. Mattson, *Energy Metabolism and Lifespan Determination*, 2003, 0-444-51492-9
15. P. Costa, *Recent Advances in Psychology and Aging*, 2003, 0-444-51495-3
16. M.P. Mattson, *Protein Phosphorylation in Aging and Age-Related Disease*, 2004, 0-444-51583-6

Printed and bound by CPI Group (UK) Ltd, Croydon, CR0 4YY

08/05/2025

01865007-0001